In Search
of the Best Strains
of Bees

Brother Adam

IN SEARCH OF THE BEST STRAINS OF BEES

AND THE
RESULTS OF THE EVALUATIONS OF
THE CROSSES AND RACES

NORTHERN BEE BOOKS ● HEBDEN BRIDGE
WEST YORKSHIRE, U.K.

DADANT & SONS ● HAMILTON, ILLINOIS, U.S.A.

Title of the German edition:
Auf der Suche nach den besten Bienenstämmen
Title of the French edition:
A la recherche des meilleures races d'abeilles

© 1983 by Brother Adam
ISBN 0 907908 06 3
British Library Catalogue No. 595.79'9 QL 568.A6
Library of Congress Catalog Card No. 83-15097

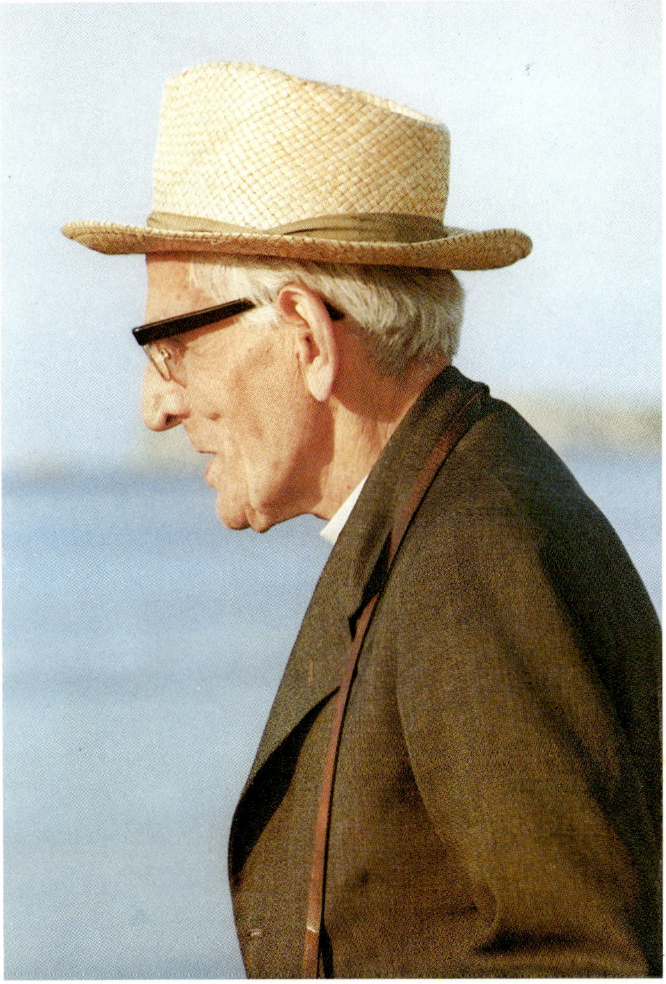

CONTENTS

Forward by Professor Dr. F. Ruttner,
   the Bee Research Institute. Oberursel, Germany 9

Author's Preface
   Br. Adam OSB, Buckfast Abbey, Devon, England 11

PART ONE: THE JOURNEYS

Map of the countries visited 16

The Purpose and Scope of this Work 17

1950 – France (23); Switzerland (26); Austria (31); Italy (36); Sicily (39); Germany (42)

1952 – North Africa: Algeria (51); Israel (61); Jordan (65); Syria and Lebanon (68); Cyprus (70); Greece and Crete (75); Yugoslavia: Slovenia (82); Ligurian Alps (86)

1959 – The Iberian Peninsula: Spain and Portugal (89)

1962 – Morocco (104); Asia Minor: Turkey (111); Turkey 1972 (119); Greece (124); Slovenia (124); Morocco-Sahara 1976 (125); Greece 1977 (132); The honeybees of Asia Minor (133); The Aegean Isles (138); Yugoslavia: Banat (143); Egypt (146); The Egyptian Bee (154); In the Libyan Desert (157); The supplementary journeys (159)

PART TWO: AN EVALUATION OF THE RACES AND CROSSES

Introductory Observations 163

The Evaluations 167

Apis mellifera ligustica 167

Apis mellifera carnica 170

Apis mellifera cecropia 175

Apis mellifera adami 178

Apis mellifera caucasica 179

Apis mellifera anatolica 180

Apis mellifera fasciata 185

Apis mellifera syriaca 186

Apis mellifera cypria 187

Apis mellifera intermissa 190

Sub-varieties of the intermissa 192
    Apis mellifera major nova 192
    The Sicilian bee 193
    The Iberian bee 193
    The French bee 195
    The Nigra 198
    The Old English bee 199
    The Heath bee 201
    North-East European and North Asian sub-varieties 202
    The Finnish bee 202
    Apis mellifera sahariensis 203

Conclusion 205

FOREWORD

A lasting contribution to progress in bee breeding has been made by Br. Adam in the past few decades. His system of beekeeping, which owes its success not only to his profound knowledge of the bee itself, but also to his remarkable organising ability and technical skill, has prompted bee breeders in more than one country to abandon outmoded methods and seek a new approach to their work.

Very early in his career Br. Adam realised clearly that with bees as with other forms of life the differences between the bees of different origin were very marked, and that it was by breeding that results of great economic value could be produced. As a young beginner he suffered heavy losses in his colonies through disease, and the development of the 'Buckfast' bee by crosses with Italians which ensued was an experience that made a lasting impression on him. But he did not rest content with his initial success. His search for material for breeding a 'perfect bee' has led him to travel in many parts of Europe and the Mediterranean countries, collecting queens and samples of bees wherever he went and noting with the practised eye of the expert the peculiarities of beekeeping in these different countries. It is indeed a pleasure to read these reports of his journeys, journeys it must be emphasised that far from being holiday jaunts were stern tests of endurance. Work on the scientific material collected still awaits completion, but it promises to mark an important step in widening, and correcting, our knowledge of the races of bees.

Br. Adam has now at his disposal a fund of knowledge acquired from his experience with the different races of bees such as is possessed by no other person. He is able to state that broadly speaking it is many of the crosses – but by no means all! – which are superior; that reciprocal crosses turn out differently; and that many *races show their true worth only when they are crossed.* The experiments he has been making these many years with the different crosses are of inestimable value for the future of bee breeding. And even when, nay rather precisely when, his judgment conflicts with the experience of others, then must it be given due attention, for it is clear that the bee is so wedded to its environment that about it there can be no judgment of absolute value, but only an assessment relative to certain given conditions. The employment of hybrids formed from different races of the bee, which with the results and successes obtained is described by Br. Adam as a way to greater honey production, seems to me to open up immense possibilities for further development. A great deal of work remains to be done, but anyone who ventures on the task in the future will have to build on Br. Adam's experiences.

This book, a mine of information for every beekeeper who is interested in bee breeding, and one of the most fascinating accounts of the travels of a bee breeder, will without any doubt occupy an exceptional position in the present day literature on beekeeping.

Oberursel, Easter
1966 F. RUTTNER

PREFACE TO THE FIRST AND TO THE SECOND EDITION

Since the appearance of the account of the final journey I made in search of breeding material, I have been repeatedly urged to publish the whole series of reports in book form. It has been pointed out to me that these reports deal with an aspect of bee-keeping which, however it is regarded, from the practical or scientific point of view, has never yet had any similar treatment in modern beekeeping literature. It is now possible for me to accede to these requests.

The purpose of this investigation and the detailed plan for its accomplishment was decided on as far back as 1948. The first journey was made in the spring of 1950, and the last one ended shortly before Christmas in 1962. The reports on these journeys appeared first in English in *Bee World*, and then in a German translation for the *Süwestdeutscher Imker*, and a French translation for *La Belgique Apicole*. Their publication now in book form will make them available to a much wider circle. The supplemontary section, giving the results of our evaluation of the various races, forms an integral part of the enquiry.

The remarks which formed a sort of preface to the first report indicated the necessity and whole purpose of these journeys of investigation. In spite of this, I soon noticed here and there that the point of the enterprise had in many instances been misunderstood. It was being assumed that I was looking for nothing other than the 'perfect bee', that is, one particular race of bees which surpassed all the others in a combination of characteristics of economic value, especially for honey production. Such a search would have been of course a fruitless venture, for Nature never breeds for the perfection of the factors we desire for our commercial needs. Nature's aim is almost exclusively for the preservation and multiplication of a type. True to this goal of hers, she breeds within certain limits, to bring about the best possible adaptation to prevailing conditions. Consequently she has bequeathed to us a very considerable number of types of bee of varying worth. As we shall see each race of bees has its own distinctive features, its good and bad traits, in each case differently emphasised and differently linked with others. The task of modern bee breeding is to ascertain which of the races has the greatest value for breeding purposes, to collect them together, to test them, and by cross-breeding combine the best of the characteristics to form new types. The realisation of these promising possibilities is the sole end of the task I set myself.

In contrast to the often groundless assumptions and baseless theories which current opinion has about different races and their hereditary descent, these journeys and investigations give one not only an appraisal of the value of the different races, but also a store of reliable information and a sort of aerial view of the interdependence of certain race groups. It is quite extraordinary how often the notions obtained mostly from hearsay about this or that race are completely contrary to fact.

It is a pity that the biometric data of the samples which I collected on these journeys is not yet available. However, my evaluation of the worth of a race relies much less on external features than on the permanently established characteristics on which in the final analysis true worth depends. In this connection theories about the origins of the primary races or about the evolution of the honeybee have hardly any bearing on the point at issue here.

These journeys too gave me a good idea of the primitive types of hive which are still in use. There is little doubt that in the foreseeable future this kind of beekeeping will be a thing of the past. In the selection of photographs for this book, I have limited myself almost exclusively to this old form of beekeeping, partly as a reminder of how it was once carried on, partly for historical reasons, for the shape and size of these hives which have often been in use from before the dawn of history provide us with a standpoint from which to form an idea of the sort of conditions in which the race of bees in question developed.

The success of journeys of this kind is dependent on the constant help and co-operation of the authorities concerned as well as the goodwill of a great number of people. It gives me great pleasure therefore to put on record here the expression of my gratitude to all those who assisted me in any way. Unfortunately I cannot possibly list all these people by name, and I have to limit myself to explicit mention of the official bodies who gave me their help. Moreover, a number of persons whose friendliness at one time or another proved so valuable, are no longer among us.

First among those who deserve mention is the English Ministry of Agriculture without whose support these journeys would simply not have been possible. Then comes the Ministries of Agriculture in Egypt (not forgetting the University of Cairo), in Israel, Greece, Portugal, Spain, Turkey and lastly in the United States of America for their very special courtesy. My grateful thanks must go the officials of these Ministries, who put themselves so generously at my service and had to share with me the rigours and hazards of travel.

Finally, I must not allow this opportunity to pass without a word of appreciation to my Abbot, the Rt. Rev. Placid Hooper, and his predecessor Abbot Bruno Fehrenbacher, who died in 1965. It was only through their understanding and their interest that the undertaking was at all feasible.

BROTHER ADAM
Buckfast Abbey, Easter 1966

*
* *

Since the publication of the first edition of this book 16 years have elapsed. Meanwhile much additional information of great value has come to light in our knowledge of the honeybee races. I was also able to carry out three further journeys – one in June 1972 to Asia Minor; another in April and May 1976 to Morocco and the Sahara; and a further one in July 1977 to Greece. The previous journeys, and the comparative tests made meanwhile on the breeding stock first collected, indicated where the best strains could be found in the respective countries. These subsequent journeys also provided the possibility of visiting regions which had to be omitted previously, due to political difficulties or lack of time.

In the concluding report of 1962 it was stressed that the task I had set myself had not as yet been finished, and that there was no prospect of a satisfactory termination in sight. In certain respects this holds good even now, for there are still extensive regions – particularly in Africa south of the Sahara – where many races of the honeybee exist, of which we have as yet little or no knowledge of their specific characteristics of possible economic value and breeding potentials. All indications point to the fact that these races are endowed with genetic possibilities of a kind not found anywhere else.

I have been well aware that I could not attempt to undertake the further journeys on my own as hitherto. Thanks to the generosity of Herr F. Fehrenbach, Ravensburg, Germany, I was given an opportunity to undertake the additional journeys. Not only did he place his car at my disposal but also shouldered the major burdens which such journeys involve. Dr. J. F. Corr, Belfast, also extended to us his support on all the journeys made since 1962, and we could not have done without his invaluable help.

Prof. Dr. A. Kirn, Reutlingen, Germany, assisted us on the search in Asia Minor; and Herr and Frau Köster, Castrop-Rauxel, on the journey to Greece. To all these people – foremost Herr Fehrenbach – I owe a boundless debt of gratitude. These journeys often proved extremely exhausting and hazardous and demanded from every participant a great measure of selfless co-operation.

The first edition of this book was sold within a very short time, and requests for a further edition came from every part of the world. However, I felt a delay was called for until the findings and results of the supplementary journeys could be included. The present edition contains all the additional information and can be regarded as an up-to-date summary of our knowledge of the honeybee races that have their habitat in the countries adjoining the Mediterranean.

Br. Adam, Spring 1982.

PART ONE

THE JOURNEYS

THE EXTENT OF THE SEARCH
A map of the countries visited

The author of this book, Brother Adam, has been engaged in keeping bees at Buckfast since 1915. The Abbey possesses some 320 colonies for honey production and about 500 nuclei for the raising of queen bees. These nuclei are situated in a secluded place in the heart of Dartmoor isolated from other bees so as to ensure the mating of queens with only select drones. This breeding station has been in continuous use since 1925. For certain special crosses a supplementary isolation apiary is used at the necessary distance from the main mating station. The innumerable cross-breeding experiments which have been made since 1925 and which are the whole raison d'etre of this undertaking would not have been possible without these facilities.

With a view to the further advancement and needs of modern beekeeping, Brother Adam undertook between 1950 and 1962 a series of journeys to investigate the precise whereabouts of the different races of the honeybee and to collect samples from the various strains found among these races. In the course of this venture, his quest took him not only to the main beekeeping centres of Europe – France, Germany, Switzerland and Austria – but also to the countries adjoining the Mediterranean – the Balkans, Italy and the Iberian Peninsula, Asia Minor, the Levant, Egypt, North Africa, Cyprus, Crete, Sicily and the islands of the Aegean. In the course of these journeys he covered about 82,000 miles by road, 7,800 by sea and 4,760 by air. The map conveys some idea of the countries visited and the distances travelled by him. The whole undertaking was financed by the Abbey.

THE PURPOSE AND SCOPE OF THIS WORK

Exactly a century has elapsed since the invention of the modern hive. With the introduction of the movable-comb hive in 1850, modern beekeeping was inaugurated. The next most important advance took place a few years later, when on July 19th 1859, the first consignment of Italian queens reached this country.

Immense strides have been made in bee culture – in the technique of managing bees, in the design of hives and appliances, and the apparatus and machinery used for the production and handling of honey. This gradual evolution has taken well nigh a hundred years. The perfection of equipment and machinery may now be regarded as a closed chapter in the development of modern bee culture. No further radical and far-reaching improvements are possible. Such discoveries and improvements as await us in the future lie in a totally different direction. It is in the bee itself that we foresee the most profound and far-reaching progress – progress of a kind that will prove almost as revolutionary as the great technical and mechanical developments that have taken place in bee culture in the last hundred years, or possibly even more so.

Apart from the invention of the movable-comb hive, the arrival of the Italian queens in 1859 has without doubt proved the most important single factor in the advancement of modern bee culture. In 1880 D. A. Jones of Canada, and in 1882 Frank Benton of the USA, visited the Middle East, to ascertain the value of the native races found in that part of the world. Though importations of Cyprian and Syrian queens were made, their efforts at finding races superior to the Italian were doomed to failure. Whilst the eastern races will never find favour, nor be able to complete with the overwhelming popularity of the Italians, they

17

nevertheless form a priceless asset for the breeding of improved strains or new races of bees.

Here in England no sustained effort has ever been made at an improvement in the bee. Intense interest is evinced in every latest innovation with regard to management, design of hives, appliances and machinery, but the infinitely more important matter, *the improvement in the bee itself*, seems to elicit no real appreciation. The current trend of economic circumstances may bring home to beekeepers the supreme necessity of an improvement in the bee. The secondary matters of bee culture, such as spring management, swarm control, etc., will then be relegated to a rank of minor significance. Indeed, with the improved bee, as visualised by us, the majority of problems that now preoccupy the minds of beekeepers will not be met with any longer. As a case of point, I will instance hereditary resistance to acarine disease. A strain that is susceptible to this disease must be treated periodically, if serious losses are to be avoided. On the other hand, a strain that is resistant will never need treatment. All the extra labour, cost of medicaments – and the inevitable losses that will occur, due to the treatment, or in spite of every remedial measure – will be completely obviated. Where resistant bees are kept, acarine, from the strictly practical point of view, does not exist. So it will come to pass in regard to many difficulties that at present cause so much extra work and concern to beekeepers everywhere.

Attempts at improvement in the bee, which have been made hitherto, are mainly in the form of line-breeding. Most valuable results can be secured by line-breeding; if carried out with patience and perseverance real progress can be achieved. But if not founded on a broad basis, and not carefully planned and executed – and especially if inbreeding is carried beyond a certain point – the results may prove disastrous. A progressive loss of vigour, as uniformity increases, precludes any far-reaching or revolutionary improvement in the bee by this mode of approach. By line-breeding, it is moreover not possible to develop a quality of which there is no trace in the genetic composition of the strain. To introduce a new characteristic, cross-breeding must be resorted to. Hybridisation is, indeed, the only possible way whereby the desirable traits of the various races can be integrated in one strain – by which means radical progress can be achieved and entirely new strains evolved.

I am not unaware of the complex problems entailed in cross-breeding the honeybee. Parthenogenesis and the haploid inheritance of the drone make the task particularly difficult, and to ensure success exceptional facilities are called for. At Buckfast we have at our disposal the indispensable prerequisites needed for this particular type of work. Obviously no selective breeding is possible without full control over the drones. This requirement is fully met in our case on nearby Dartmoor, where we have had a mating station in operation since 1925. The many years of experimental hybridisation have also provided the experience on which the successful execution of a project of this kind ultimately depends, and has in addition enabled us to visualise the immense potentialities cross-breeding can offer.

In the case of the bee, *the best possible breeding stock* forms the most vital requirement. To use second-rate stock for cross-breeding would with infallible certainty invite untold labour, losses and disappointments. Queens imported through the ordinary trade channels are valueless for work of such an exacting nature. To get hold of the right type of breeding stock I felt I had no other choice than to make a personal search of the native habitat of the races required for our breeding experiments. In addition, as each race requires a great number of strains, of widely varying value, the selection could in every case only be made on the spot. Furthermore, strains suitable for hybridisation are generally only found in remote isolated country districts, where in complete seclusion the purity of the race has been preserved since ancient times, and where by continued inbreeding the greatest genetic uniformity has been attained. Breeding stock of such a high standard is clearly unobtainable by any other means.

I have therefore undertaken a series of journeys, which will eventually cover all the countries bordering the Mediterranean that possess an indigenous bee of outstanding merit. As already stated, the primary purpose of this search is the collection of the best possible breeding stock which we require for hybridisation. However, there is a series of secondary objectives, all of which exercise a direct influence on the ultimate success of the task we have set ourselves.

One of the most important secondary objectives of my search is to secure first-hand information of the range of variation of the distinctive morphological and physiological characteristics of each geographical race. Out literature contains extremely little reliable information – if any at all – in regard to this matter, which might be of practical value to us. An exact and comprehensive knowledge of the range of hereditary characteristics of the bee is an elementary requisite to success in cross-breeding. These journeys provide an inestimable fund of knowledge, of the specific kind needed, which could never be obtained in any other way.

In addition, samples are collected on behalf of Rothamsted Experimental Station of every race and distinctive strain for biometric purposes. Each sample comprises about a 100 bees, which are preserved in bottles containing a special medium. The data obtained from these samples will form a permanent record for future reference.

It may not perhaps be generally realised that many races and strains are gradually but surely approaching extinction, due to widespread hybridisation. Indeed, according to my findings, this regrettable state of affairs has already progressed so far that the native bee, in its original purity, does not exist any longer in a number of countries – or if still extant, it is only found in remote parts of isolated valleys, shut off from general intercourse. This holds good at least in respect of Western Europe. From the genetical point of view this is a most deplorable development, for many valuable characteristics have been submerged or already completely lost in the welter of indiscriminate hybridisation. Mongrels are of no value whatever for breeding. Therefore no specimens now collected will form an invaluable record for reference in the future – quite apart from any immediate gain the biometric studies will reveal.

In my work on the Continent I took a particular interest in every effort directed towards improvement in the bee. An immense amount of work has been done in this direction, of which we here in England possess only a very hazy notion. The great movement – Die Rassenzucht – was inaugurated in 1898 in Switzerland by Dr. U. Kramer, and mating stations have been in use for well nigh half a century in Austria, Germany and Switzerland.

Last, but not least, I realised that a visit to the Continental Research Institutes, and the setting up of a direct link with the leading scientists and beekeepers abroad, would prove in many ways an invaluable help in the task we have set ourselves. An exchange of views, and a verbal discussion of the many complex problems, invariably leads to a better mutual understanding and appreciation of the real nature of the difficulties at stake.

Beehouses fitted with skeps, as this one in the French Jura, was the only form of beekeeping known a hundred years ago: they are now on the verge of extinction.

A migratory apiary in the Corbieres, where the world-famous 'Narbonne' or Rosemary honey is produced.

In the Jura: the French bee demands a large comb and a roomy hive, such as the 12-frame Dadant pattern used here.

Erica arborea, indigenous to the Mediterranean countries and the southern Black Sea coast, gives a water-white honey – and also the material from which briar pipes are made.

PLATE 1

FRANCE

A primitive apiary in the Le Gatinais area. Apiaries such as this at one time formed part of the normal English country scene.

22

1950
FRANCE – SWITZERLAND – AUSTRIA
ITALY – SICILY – GERMANY

The first journey made by Br. Adam was carried out in two sections. On March 20th 1950 he set out from Newhaven for the Continent and then continued his journey by car to Southern France. At the conclusion of his task in that country he spent April and May in Switzerland and Austria, including two brief visits to South Germany. At the beginning of June he returned to England for three months, to attend to various urgent tasks at Buckfast. On August 21st he set out for Austria again, to complete the search in Carinthia and Styria, before proceeding to Italy and Sicily. On the return journey the French-speaking section of Switzerland was visited as well as the principal research stations in Germany.

FRANCE

On my arrival in France I immediately made my way to the south for obvious reasons. Spring had made its entry in the extreme south. In fact the main honey-flow was already in progress at that time on the Mediterranean seaboard. In the Corbières and Provence rosemary was in full bloom. Indeed, the season was already well advanced. At Ceret, not far from Perpignan, white clover was in bloom along the roadside on March 28th. The Corbières – one of the most marvellous nectar-secreting regions in the world – is situated between Narbonne and Perpignan to the east, and Carcassonne and Quillan in the west. The world-famous Narbonne honey is derived from the rosemary of the Corbières. Rosemary thrives best in this hilly, rocky and apparently barren country. A net gain of 15 lb. per colony in a day is no uncommon event when rosemary is in full bloom, and when climatic conditions are just right for best results. Unfortunately in this region of France a wind of gale force – on about 220 days in a year – all too often spoils a beekeeper's expectations. The westerly currents of air are diverted into this area by the Pyrennees to the south and by France's Central Massif to the north. As the air currents reach the narrowest gap, near the Mediterranean coast, the wind in the Corbières often attains a velocity of 85 miles per hour. Though the sun may shine, bees cannot brave a fierce wind. It will be readily realised that bees of an extremely hardy type, endowed with exceptional stamina and wing power, are a vital necessity in this region of France. And it is therefore not surprising that some of the best strains of pure French native bees are still found in the Corbières. The pure French bee, as we knew it 20 or 30 years ago, is nearing extinction. Only a comparatively small number of commercial beekeepers still favour the native black bee. And these few experience almost insuperable difficulties in keeping their strains pure, because of the widespread use of hybrids. The vast majority of French beekeepers make use of American-bred Italian queens crossed with local drones. The progeny of the pure American queens appears valueless for the production of honey, but a first-cross gives very satisfactory returns. With the exception of a few isolated instances, I found nothing but first-cross hybrids or mongrels wherever I went in France. One of the best breeder queens that we at any time possessed came from the Le Gatinais district. On my search of this area, in May 1950, I could not trace a single colony of the pure native bee. Instead I came across some of the most horrible mongrels I have ever seen.

    The decline of the French native bee is no doubt largely due to its vile temper. When thoroughly roused, especially towards the end of the season, or immediately the crop has been taken off the hives, it will sting every living creature within reach. It is also inclined to swarm unduly, and to gather propolis in excessive amounts. Indeed, in this respect it excels any other bee I know, excepting the Caucasian. The interior of some of the hives I came across in France was simply plastered with propolis of the sticky resinous type, which makes the handling of combs an arduous and unpleasant task. In spite of these rather serious defects, it would be an irreparable misfortune and loss if the native bee of France succumbed to the current trend of indiscriminate hybridisation. Its defects are great indeed, but the good qualities it possesses are as great. It is very hardy, long-lived, strong on the wing and one of the best foragers. It is also a good comb builder, and the cappings produced by some of the strains are almost faultless.

The French bee may be regarded as a near relation of the Central European brown bee, with the difference that many of the good and bad characteristics of the latter are developed in the former to extremes. From the breeder's point of view, it is the more valuable of the two, for the French bee lends itself particularly well to cross-breeding. Bad temper, however pronounced, can be readily eliminated in the subsequent segregation and recombinations of characteristics.

The French native bee is afflicted by one further defect, a defect manifested by nearly every variety of the Central European brown bee: a pronounced innate susceptibility to brood diseases. Here again, in the French bee the susceptibility is manifested in a much more accentuated form than in any of the other related varieties of the brown bee, as a matter of fact almost to the same degree as the correlated defect, namely a lack of cleanliness or tolerance of anything abnormal within the brood nest, which is one of the predisposing causes to diseases affecting the brood. The quite remarkable tolerance of wax moth of the French bee is clear evidence of a deficient housekeeping ability.

I was told that beekeeping has been on the decline in France during the past 150 years. However, there are definite indications of a revival, and commercial beekeeping is at present carried out on a far more extensive scale than here in England. A country that possesses such a diversity of nectar-bearing flora – where sainfoin is found wild by the roadside and on every bit of waste ground, and where rosemary, lavender, buckwheat and heather abound – bee culture should flourish. Indeed, I believe France has at its disposal some of the very best beekeeping territory in Europe. The methods of management in vogue amongst commercial honey producers cannot be described by our standards as intensive. Notwithstanding this, good crops are secured. But it would seem to me that a more intensive system of management would render the keeping of bees still more profitable.

SWITZERLAND

I visited Switzerland on no less than three different occasions in the course of my travels in 1950. The first visit took place early in April. The weather was still very wintery at the time, and it was clearly too early in the season to devote any attention to the bees. Therefore on arrival at Berne my first steps were directed towards the Mecca of beekeeping research – the Liebefeld Institute. Professor Dr. O. Morgenthaler was unfortunately absent; however, Fräulein Dr. A. Maurizio, whom I first met at the International Congress in Amsterdam, kindly introduced me to the staff. She also explained in detail her own work of pollen analysis, a subject on which she is everywhere recognised as the leading authority. It was in her Department that I first tasted that very choice and delicious honey obtained from the *Alpenrosen*, a dwarf species of rhododendron which only thrives at great altitudes and in non-calcareous regions of the Alps. In my estimation, it is the loveliest honey produced on the Continent or possibly in the world. In the laboratory of Herr Schneider und Brügger we discussed the many problems associated with acarine disease, and the latest research work carried out by them at Liebefeld. In Switzerland great efforts are made to combat this menace by restrictions on the movement of colonies in areas where the disease is found, and by compulsory treatment of every stock in a district where acarine is confirmed. By these countermeasures it is hoped the disease will be eradicated, or anyway that the losses will be reduced to a minimum. Unfortunately, in the countries surrounding Switzerland, acarine is spreading its tentacles ever and ever further afield. In my estimation, treatment can never prove the final solution to this problem. For the present at least, acarine has not reached the virulence on the Continent which it assumed here in England when the disease attained its peak of devastion.

In this and subsequent visits to Liebefeld I was particularly interested in the work of Herr W. Fyg. The work carried out by this eminent zoologist is not as well known here in England as it deserves. Herr Fyg is very modest, shy and diffident of his great abilities. He is unquestionably the most outstanding authority on all matters relating to the structure and the many pathological conditions affecting the reproductive organs of the queen bee. As far as I know, he is the only scientist in the world who has made the study of these problems his exclusive task. His contributions to our knowledge of the anatomy, physiology and pathology of the queen bee are indeed invaluable. I was deeply interested in this particular sphere of research, for it threw much light on many problems which had hitherto eluded a satisfactory explanation.

Due to arranagements made previously with the beekeepers of Carinthia, this first stay in Switzerland had to be of short duration. However, I returned to Berne on May 15th. My main task on this visit was confined to a study of the various distinctive strains of the native bee, which have been developed in Switzerland in the course of the past 50 years. Dr. Morganthaler and A. Lehmann kindly organised and accompanied me on these journeys. Dr. M. Hunkeler, *Chef der Rassenzucht,* joined us on the second day.

There are a great many strains, developed from the common native bee, in use in Switzerland. It is generally assumed that each and every one of these strains embodies some special characteristics, or some inherent specific adaptation to the particular environment where it originated. It is therefore further assumed that the best possible results can only be obtained from a strain which has adjusted itself fully to the modifications imposed by each immediate environment. It would serve no useful purpose to describe these strains in detail. However, there is one really outstanding strain, which deserves special mention, the 'Nigra'. This strain is in fact a true Swiss creation. It was founded about 50 years ago by F. Kreyenbühl. Up to some years before the last war it was the most favoured and widely used strain in Central Europe. It has fallen into disfavour in Germany during the last ten years, and is now being rapidly superseded by other varieties. Faulty breeding, and too great an attention to external characteristics, have probably brought about a degeneration of the 'Nigra' bred in Germany. We have given this strain an extensive trial in our own apiaries and we were greatly impressed by it. It embodies many desirable qualities, but unfortunately it has one serious defect, which completely dissipates all its good traits: it swarms to excess, particularly when crossed. For this reason it does not lend itself, in our climate, for commercial honey productions. Were it not for this one defect, there would be much to commend the 'Nigra'. As its name indicates, it is black – not brown. Indeed, it is jet black. The extraordinary colour, the unusual swarming tendency, and some other traits in the character of the 'Nigra' bee seem to me to indicate a close affinity to the German heath *(Apis mellifera var. lehzeni)*.

Apart from the main objective of my work in Switzerland, I gained first-hand experience of Swiss methods of beekeeping, and the technique of handling colonies in house-apiaries. A beehouse has undoubted advantages. But a beehouse will never lend itself to the speedy operations and manipulations, which are the *sine qua non* demanded by the most advanced methods of bee management. The Swiss have admittedly developed an extraordinary skill in handling combs, in their withdrawal and replacement, by the use of the special tongs, required for this purpose. Nevertheless, quite apart from any practical considerations, in regard to the physical impossibility of executing every operation and manipulation with the utmost speed and efficiency, a beehouse seems in many cases to possess other additional drawbacks. The excessive protection, and the excessive heat developed in these heavily-timbered buildings during the summer months is not – according to my experience – conducive to a normal, natural and healthy colony development. I certainly went away with the very definite impression, that in the beehouses I visited the bees were kept far too warm to ensure the best results. It should not be presumed that house-apiaries are in use throughout Switzerland. In the French-speaking zone of western Switzerland, the hives are set out in the open, just as is done here in England.

A typical Swiss beehouse in the Bernese Oberland.

Br. Adam in conversation with Professor Dr. O. Morgenthaler, founder of the Bee Institute at Liebefeld and first General Secretary of Apimondia.

Professor Dr. L. Armbruster, Lindau, editor of *Archiv für Bienenkunde*.

SWITZERLAND

PLATE II

GERMANY

Lüneberger Heide, Germany: A type of beekeeping that will soon be a thing of the past.

At Erlangen: Dr. K. Böttcher and Br. Adam

I returned once more to Berne on October 8th. On this occasion my quest took me to the extreme western sector of Switzerland – the Neuchâtel area. The Dadant hive is used almost exclusively throughout the French-speaking zone of Switzerland. Indeed, the language demarcation zone seems to form the actual dividing line between two totally different systems of beekeeping. In the German-speaking section beehouses are in general use and a size of frame of approximately equal comb area to the British Standard; in the French-speaking part of the country the large Dadant hive is seen everywhere.

The organisation of the German-Swiss Beekeepers' Association is the most advanced and in many ways the most progressive of its kind in the world. Their bee-disease insurance scheme, their honey control, and above all the improvement of their native bee by means of controlled mating of queens – the work initiated by Dr. U. Kramer in 1898 – are some of its most outstanding achievements. In 1950 the Association had no less than 183 mating stations in operation.

In spite of the immense achievements attained, I could not fully convince myself that by the type of bee in use, and the system of beekeeping in vogue, the maximum return per colony was in fact obtained in Switzerland. So many of the arguments brought forward in favour of the native bee, and the particular system of beekeeping, reminded me of the views and opinions held here in England 35 years ago. With extreme tenacity some of our leading men then maintained that the old English native must – *ipso facto* – be the best bee for our climate. It was argued with some justice that, in the course of thousands of years, natural selection would with unfailing certainty evolve and mould a bee best adapted to the peculiar requirements of our island climate. But I know from the hard lessons of first-hand experience how utterly fallacious this argument proved. However, on my journeys on the Continent I was again and again involuntarily reminded of the fallaciousness of such reasoning and the consequences it will lead to. It is so very easy in beekeeping to stray into a blind alley – with the added difficulty that one is often in ignorance that false theoretical considerations have led inevitably into a cul-de-sac. If the bees fail to build up in spring, or if they fail to do well at any time of the season, it is to easy – all too easy in fact – to believe with the utmost conviction that the weather has been at fault, or that due to some inexplicable cause the flowers failed to secrete, or anyway failed to secrete as well as they should have done. On the American Continent beekeeping is guided too much by purely commercial and practical considerations. In Central Europe the reverse is the case: abstract considerations tend to override all practical aspects of beekeeping. Theoretical advantages and drawbacks, when put to the 'acid test' of severe practical beekeeping, prove all too often fictious.

My visits to Switzerland have been all too brief. Among such a highly organised community of beekeepers there would have been so much to learn.

AUSTRIA

There are three distinct varieties of bees found in Austria: the Central European brown bee; the Alpine bee; and the Carniolan or Carnica bee. The brown bee is found in Upper Austria; the Alpine variety is confined to the northern regions of the Alps, mainly in the valleys of the Salzach and Enns. The high mountain range, known as the Hohe-Tauern and Niedere-Tauern, forms its southern boundary. The native habitat of the Carniolan bee is immediately south of the Tauern, in Carinthia and Carniola. The Dolomites to the west, and the Carnic Alps to the south-west and south, form its main boundary. The range of its geographical distribution to the north-east, east and south-east has not as yet been determined with certainty.

I have examined a number of strains of the Alpine variety. In many respects they seem identical with those found in Switzerland. As far as I could ascertain, all these Alpine strains are merely forms of the European brown bee, with a few slight variations, or modifications brought about by the natural isolation of the mountainous type of country. Whilst both the Swiss and Tyrolese are some of the best bred strains extant, they do not embody any characteristics of outstanding merit. With the exception of the 'Nigra', they have a close similarity to the old English native bee. Indeed, these Alpine strains are in reality the only remaining closely related representatives of the original Central European brown bee, which in its true form – with very few exceptions – must for all practical purposes be regarded as extinct.

On the Continent the Carniolan bee is commonly known as the Carnica. In the English-speaking countries it has been given the name Carniolan, for the majority of importations in the past came from Carniola (or Krain), which before the First World War formed a Province of the Austrian Empire, but was incorporated in Yugoslavia in 1919. Carniola may perhaps be the geographical centre of the native habitat of this bee, but Carinthia is undoubtedly one of the main centres of its distribution. Moreover, the high mountain barriers that encompass Carinthia have effectively preserved the purity of this race – as perhaps nowhere else within its sphere of distribution – from time immemorial. The deep seclusion of the valleys, the well-nigh inaccessibility of many Alpine farmsteads, the severity of the climate and scarcity of nectar-bearing flora, have further contrived to bring into being, in Carinthia itself, many widely distinctive strains of this bee. Natural isolation and natural selection have here worked harmoniously hand in hand to the development of these distinctive strains. Carinthia and north-western Yugoslavia represent a veritable El Dorado to the enterprising student and breeder of Carniolan bees. This race has been described by one authority as a greyish-black version of the yellow Italian bee. Except for colour and the grey overhair, the Carniolan approximates to the Italian more closely than any other race. However, the true Carniolan is, without any doubt, a distinctive sub-species of *Apis mellifera*. But the range of variation, between the various sub-varieties and strains, is probably greater than in any other race – according to our present knowledge. Leaving out of consideration morphological differences, the variation of physiological distinctions between one strain and another is very pronounced.

One of the most outstanding good traits of the Carniolan is its extraordinary docility. As far as my experiences would indicate, it is one of the most docile of races. It can be handled with absolute impunity, without veil or any sort of protection. The bees keep calm and bear manipulation with the greatest composure, though if required they are easily shaken off the combs, and in this respect behave quite unlike the Italians. However, in the course of my quest I came across some strains that could with justice be described as bad tempered. Of the other good qualities this race possesses I must mention its incomparable hardiness, longevity, and wing power. The long, drawn-out winters, the extreme cold, and the general severity and changeability of an Alpine climate, as well as the scanty nectar-bearing flora, have contributed from time immemorial to the development of these characteristics. The extraordinary stamina of the Carniolan verges on the unbelievable. For instance, on April 19th, I visited an isolated farmstead, situated at a height of about 4,200 ft. on a bare bleak mountainside. The weather at the time was bitterly cold, and the mountains around were blanketed in snow. The bees at this farm were kept in typical Carinthian box hives. These box hives are approximately 6 in. high, 10 in. wide and 39 in. long, and are constructed throughout of $\frac{3}{8}$ in. timber. To comply with tradition, the owner of this farm would never winter more than eight colonies. So in this case the boxes were stacked in two tiers of three, and one of two, side by side. Beyond the protection this mode of stacking afforded, the thickness of the timber of which the boxes were constructed, and a makeshift roof to keep off the rain, these eight colonies had virtually no further protection or shelter. Yet, when the front of each of these box hives was removed, without the use of any smoke or subjugating material, every box was full of bees. Indeed, some of the colonies formed a cluster 3 in. in depth in front of the combs built the previous year.

    The Carniolan forms small colonies in the autumn and, consequently, manages to winter on a minimum of stores – a very desirable and valuable trait – in contradiction to the Italian. I was assured that colonies will come through the winter on 6 lb. of stores and, it must be remembered, stores at that not of the best quality. Bees in Carinthia winter mainly on honeydew honey obtained from conifers, or buckwheat, if colonies are moved to the buckwheat regions north-east of Klagenfurt. But the point is that Carniolan colonies build up very rapidly in spring, as soon as pollen is available from *Erica carnea* and the wild crocus. The former commences to bloom about mid-March, and the latter about mid-April. But the weather at this period of the year is most fickle in Carinthia, as I witnessed myself. It may turn very warm suddenly, oppressively warm in fact. Then next day there may be a return to mid-winter conditions. In such a variable spring, bees must necessarily be endowed with an exceptional power of adaptability and endurance. Spring dwindling, in such wide variations in temperatures, would spell the doom of a colony.

The Carniolan is regarded by many authorities as the honey producer *par excellence*. There are, indeed, many cases on record where it has done supremely well, especially when crossed. As a matter of fact, a first-cross is reputed to hold the world's record of surplus ever obtained from one colony. All the evidence seems to indicate that a good strain embodies the requisite faculties that go towards the making of a superb honey producer. But according to our experience, there is a wide difference between the various strains, and the best is not found by the wayside.

The Carniolan possesses an exceptional tongue-reach which is of particular importance where red clover is cultivated extensively. She is, in addition, a good comb builder and tends to cap with paperwhite cappings. But the cappings are rather flat, not convex. The true Carniolan collects less propolis than any other European race; it uses wax rather than propolis for closing crevices in the hive. I regard this as a most desirable quality, for nothing makes manipulation of combs a more unpleasant task than an excessive use of propolis, especially when it is of the sticky, resinous kind. But not every strain found in Carinthia will produce white cappings or utilise a minimum of propolis.

The one overriding defect of this race is its excessive swarming propensity. A race or strain that is given to swarming inordinately is of no practical value, here in England, for the commercial production of honey. Every desirable quality is frittered away, is dissipated by this one defect. Recently we tried out queens obtained from four different Continental breeders.

In Carinthia wild crocus provides an abundance of pollen in early spring.

Ost-Tyrol – A log beehouse in a remote part of the Virgental.

PLATE III

AUSTRIA

A charming head-piece of a Carinthian box hive depicting a well-known episode in the life of St. Isidore.

The Rosental, Carinthia, bordering on Yugoslavia: the present-day centre of dealers in Carniolan bees.

Carinthian box hives on a remote farmstead situated at 4,200 ft. on the Katschberg Pass.

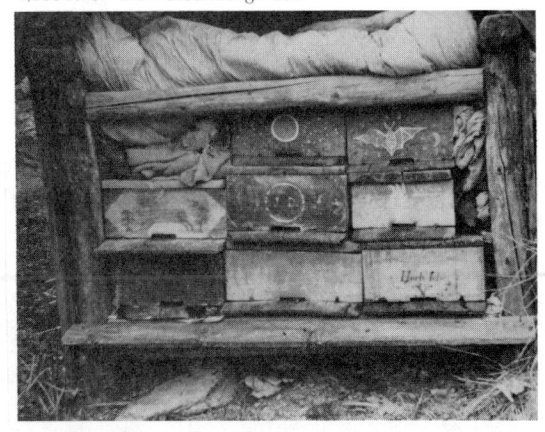

35

Every one of these commercial strains proved valueless in our apiaries because of this uncontrollable swarming tendency. It must be realised that, until recent times, this trait has been deliberately fostered in Carinthia, and is still fostered wherever the primitive box hives are in use. However, it can presumably be eliminated, or anyway reduced within tolerable limits, by selection. The bee imported from Carniola 40 or 50 years ago was far less disposed to swarming than the more recent importations. Though I believe I have found one or two strains on last year's quest, which may meet our requirements, only an actual test in our apiaries will determine the matter. According to the information placed at my disposal, which my own observations tend to confirm, there is every indication that the indigenous bee of Carniola and the adjoining territory further south is not identical in many respects with the strains found in Carinthia. I hope I shall be able to settle this important question in the near future.

I have not made any attempt at a detailed description of the less obvious characteristics of the Carniolan bee. This is clearly beyond the scope of this account. Moreover, it is in many ways a mystery race, the scope and possibilities of which have not as yet been fully fathomed, for many of its hereditary potentialities lie dormant and only come to light in cross-breeding.

ITALY

According to my timetable I intended to spend a month on Italian territory, of which a week was set aside for the exploration of Sicily. A great deal of ground had to be covered, and as the journey progressed it became apparent that the full programme could not be carried out. I was compelled to omit the north-eastern section of Italy, the territory adjoining Yugoslavia. It is in this zone that the Italian and the Carniolan bees have intermingled from time immemorial, and where in all probability fixed intermediate strains, embodying the desirable characteristics of both races, will be found. A fixed strain of this type would be an invaluable asset. I now hope to include this area in conjunction with a visit to Yugoslavia.

The world-wide overwhelming popularity of the Italian bee is beyond dispute. Indeed, I believe apiculture would never have made the progress it did without the Italian bee. Commercial beekeeping, as at present carried out in every large honey-producing country, would be well nigh a practical impossibility without it. The Italian bee is one of Nature's gifts, bestowed on a land endowed with an unparalleled lavishness of blessings. It is not a perfect bee by any means, but Nature has furnished it with a combination of desirable qualities in a measure not possessed by any other race. The Italian bee has its defects – serious defects in fact – which have denied it an absolute universal popularity. Its general characteristics are so well known that it would serve no purpose to mention them here. But the principal defects, as I see them, need pointing out. The Italian been tends to breed to excess at the conclusion of the main honey-flow, and with few exeptions it is extravagant with stores. It is not thrifty and lacks the hardiness, stamina, longevity and wing power manifested in varying degrees by most of the other races. It drifts badly; and because of this lack of stamina it is subject to spring dwindling, whenever climatic conditions hamper colony development early in the season.

There are, according to my findings, three distinct varieties of the Italian bee: the dark leather-coloured variety; the bright yellow kind, as usually supplied by the commercial breeders; and a very pale lemon-coloured type, not often seen. The so-called Golden Italian is not a true Italian bee at all. It is an outcome of a cross between the Italian and a black bee, as our cross-breeding experiments have clearly demonstrated.

Experience has shown that the leather-coloured bee surpasses in economic value the more attractive bright yellow variety. The first queens exported from Italy came from the Ligurian Alps – hence the name Ligurian bee. These original importations, made nearly a 100 years ago, were mainly of the leather-coloured variety, and it was undoubtedly the tawny Ligurian native bee that established the reputation of the Italian. According to my findings, the true leather-coloured bee is found only in the Ligurian Alps, in the mountainous region between La Spezia and Genoa. Immediately west of Genoa, hybrids make their appearance. In the region of Imperia and San Remo the French black bee – with its characteristic spitefulness – encroaches on Italian territory. In my estimation the tawny Ligurian bee, as found by me in the Ligurian mountains, embodies in an outstanding measure all the good qualities which have made the Italian so popular throughout the world.

The geographic distribution of the bright yellow variety is in the north of Italy mainly confined to the Plain of Lombardy, and southwards to the entire Peninsula as far as Catanzaro. South of this region, the worst possible type of mongrel extant dominates the remainder of Calabria – a heterogeneous conglomeration of the yellow Italian and black native bee of Sicily. Though I was not able to explore entirely the regions north of the Plain of Lombardy, yet judging from the sectors visited and the information I obtained, hybrids predominate in the southern foothills of the Helvetic Alps. In the Dolomites and the region around Bolzano, the prevalence of hybrids is very marked. On the other hand, around Lake Como and the adjoining Tessin, the yellow Italian is found more commonly. But black bees and mongrels clearly dominate the area further west, in the valley of Aosta, as I ascertained. In the regions where the bright yellow Italian bee is indigenous, the tendency to propolise increased progressively the further south the search took me.

I called on most of the noted breeders of Italian queens near Bologna. They assured me that their customers demanded bright yellow queens. Every trader supplies what their customers demand – he must necessarily do so if he is to remain in business. Undoubtedly these queen-rearing establishments around Bologna supply the best strains procurable of the bright yellow type. But I have not the slightest doubt that the tawny Ligurian is by far the better bee from the strictly practical point of view.

There are relatively few commercial beekeepers in Italy. It is mainly a country for small-scale beekeeping. Agriculture is carried out too intensively to permit the keeping of a large number of colonies in one locality. For this reason the mountain areas where wild thyme, sage, heather, etc., abound, offer the most favourable prospects for the commercial production of honey. Calabria seems particularly favoured in this respect. There are, in addition, extensive orange groves and lemon plantations along the coast, which offer a bounteous crop in early spring before the mountain flora comes into bloom. The mountains of Calabria must be a lovely sight at the time when the Mediterranean heath *(Erica arborea)* is in full bloom. I was told that this species yields a honey water-white in colour, which extracts by centrifugal force. It would seem that in southern Calabria bees often secure part of their winter stores from figs; that is, they gather the juice of over-ripe figs. The fruit abounds in this region, and the second crop (because of the small size of the figs) is usually not harvested. One beekeeper assured me that his colonies made a net gain of 15 lb. from this source the previous autumn. Bees winter seemingly perfectly well on the juice of figs in this sub-tropical climate.

Apiculture has not as yet reached a high level of efficiency in Italy. However, there are indications of an awakening. The size of frame in common use is the Dadant and Langstroth. Log hives (boxes about 10 in. square and 24 in. high) are still widely used in Campania, in the Alban hills, and north-west Italy.

SICILY

On my arrival at Messina, on September 19th, I made my way in company of M. Alber to Randazzo, situated on the northern slope of Mount Etna. This is a noted area for beekeeping. Not so many years ago, only the native bee of Sicily *(Apis mellifera* var. *sicula)* could be found throughout the island; but in recent years queens from northern Italy have been widely imported. Whether these importations of the bright yellow bee will eventually prove to the good of apiculture in Sicily is at the moment an open question. A well-known authority in Rome expressed concern and grave doubts regarding these importations. In the neighbourhood of Randazzo, I could find nothing but hybrids. From the information placed at my disposal, this seems to hold good throughout the north-eastern section of the island. However, the main purpose of my visit to Randazzo was to gain an interview with

SICILY

An apiary near Noto. Ferula hives were used here, but some of identical size were made of wood with movable combs.

PLATE IV

ITALY

One of the many commercial queen-rearing yards in the Bologna region.

Old-time box hives can still be found in the more remote parts of Italy.

Cavaliere P. A. Vagliasindi, the foremost authority on beekeeping in Sicily. On his advice we left for the extreme south-eastern sector of Sicily, for Noto and Ragusa. It is in this region that the carob tree *(Ceratonia siliqua)* grows more abundantly, and where a vast influx of colonies takes place annually, from the neighbouring hills, when the carob comes into bloom in October. The carob, seemingly, is one of the most profuse sources of nectar. The trees were in bud at the time of my visit, but the migration of the beekeepers from the hills had unfortunately not yet begun. I thereby missed an opportunity of gaining a better knowledge of the range of variation in the characteristics of the pure Sicilian bee. However, I managed to secure a few queens in this area of the pure Sicilian bee.

The *Sicilian* is reputed to be a near relative of the Tunisian bee. But as far as I know this question has not been determined definitely. At the time I called in Sicily, it was very difficult to form a clear idea of the general characteristics of the Sicilian bee. My visit coincided with the conclusion of the long drought of summer, and the autumn rains had not yet set in, nor had the main flow from the carob yet begun. All the colonies were therefore at their lowest ebb of strength. Practically no brood was present in any of the colonies examined and every one was nearly destitute of stores. However, I conclude that the native bee of Sicily must posses great stamina, and it must be very long-lived; it could not otherwise survive the long periods of dearth. It has the reputation of being bad tempered. I was able to handle them without any protection – at least the colonies I examined in the Noto and Ragusa districts. On the other hand, in central Sicily I came across some frightfully vicious colonies. I was assured that the Sicilian bee is disinclined to robbing, or it would seem does not rob at all – which, if true, would be a most valuable trait. Only an actual test of this race in our own apiaries will reveal the true characteristics of this bee, and determine is general usefulness.

In many parts of Sicily beekeeping is to this day carried out in as primitive a manner, for all we know, as in remote antiquity. Movable comb hives are found here and there, but the majority of colonies are accommodated in box hives with fixed combs. The material of which these boxes are constructed is either wood or, more generally, the stems of a giant fennel *Ferula thyrsifolia* – hence the name ferula hives. The stems are about 1½ in. in diameter, and are extremely light – indeed as light as cork, and no doubt of equal insulating value. Whether constructed of wood or *Ferula* stems, these box hives are about 10 in. square and 30 in. long. Both ends are closed by a tightly fitting board. The space occupied by the bees can be contracted, whenever necessary, by pushing the board at the back further towards the centre of the hive. These boxes are invariably stacked in tiers, usually five on top of each other, and as many as 20 hives may be found side by side forming one huge block. An open shed, constructed of stone and with a sloping roof made of tiles, affords the necessary protection from sun and rain. All manipulations are effected from the front; the boxes are slipped in and out as required. When the honey is taken, the bees are not destroyed, but merely driven into the forward compartment by means of smoke. The ferula hive is typical of Sicily, and as far as I know is not found anywhere else in Europe.

The nectar-bearing flora of Sicily is largely sub-tropical. The main sources are lemons, oranges and mandarines, acacia, carob and mountain thyme, together with a number of other minor sources.

GERMANY

When on my way to Austria, on April 12th, my route took me through Lindau, situated on the eastern fringe of the Lake of Constance. Here lives in enforced retirement one of the foremost leaders in apicultural science in the world – Professor Dr. L. Armbruster, at one time Director of the Bee Research Institute at Berlin-Dahlem, and Editor of the exclusive *Archiv für Bienenkunde*. I say 'enforced retirement' for he was compelled to surrender his directorship because he dared to say 'No' to Nazi instructions, when Hitler took over in 1933. Prof. Armbruster is a man of high conscience and a fearless defender of truth. Unlike others, he had the force of character to brave dismissal, and poverty, rather than deny the permanent and higher values of life.

There are relatively few outstanding leaders in apiculture. Scientists of many nations have made valuable contributions to the general knowledge of bee culture, but their discoveries are usually confined to a specialised sphere of research. By leaders I mean men endowed with a thorough grasp of the fundamentals of beekeeping – men of wide vision and judgment, who are able to cut a clear path through the tangle of purely theoretical considerations and prejudices, and who will not allow themselves to be led into barren deserts of a pseudo-scientific bee culture. In the strictly practical field of commercial beekeeping I regard E. W. Alexander of Delanson, USA, and R. F. Holterman of Brantford, Canada, as the leaders who gave us the most valuable information. In the scientific and theoretical sphere – especially in all matters relating to heredity and the application of the Mendelian laws to the breeding of the honeybee – the writings of Prof. Armbruster have given to me personally the most valuable guidance and inspiration. His *Bienenzüchtungskunde*, published in 1919, gave the key to the development of our strain. Amongst the welter of confusion and ideas expressed in the German literature on this subject, Prof. Armbruster's views proved a clear beacon of light indicating a path towards the goal I had set myself. His aim and my aim in breeding are identical: the development of a strain that will produce the *maximum returns with a minimum of labour*. He maintains, and so do I, that this objective can only be attained by cross-breeding, by combining in one strain, as far as is possible, the desirable characteristics of the various geographical races. Nature can never bring about such a combination; it can only be effected by the direct intervention of man. I well realise that these views and tenets are in absolute conflict with the generally accepted teaching on the Continent.

Apart from his writings on the specialised subject connected with the science of bee breeding. Prof. Armbruster has rendered apiculture throughout the world an incomparable service by the publication of the *Archiv für Bienenkunde*. The *Archiv* is the only journal that deals with every scientific aspect relating to bee culture. Everyday practical problems are discussed in its pages as well, but objectively and from a highly scientific point of view. The strictly practical application – and implication – of every proposition put forward is never lost sight of. Prof. Armbruster is too much of a realist to be led astray in mere theories. Nevertheless, to a reader not particularly scientifically-minded, much of the material may seem abstruse.

I had never met Prof. Armbruster before. Therefore, in passing through Lindau, I seized this opportunity to make his personal acquaintance. He at once very generously offered to render me any help he could. In the course of my work, I called on him on no less than four occasions in 1950, and each time I left with a wealth of information. On my subsequent return from Austria, he acted as my guide in the exploration of the adjoining territory of Lindau, known as the Algäu. This section of South Germany is in many ways very similar to South Devon – climate, rainfall and flora. At the time of my visit the dandelions transformed the meadows into a carpet of gold. The yield from this source is often prodigious. There is a case on record of a colony making a net gain of 16½ lb. in one day. This is a truly astonishing performance, considering the earliness of the season. The record was confirmed by Prof. Armbruster. On this occasion I also visited one of the foremost commercial beekeeping concerns in Germany, Firma Mack of Illertissen. This firm operates a thousand colonies. Migratory beekeeping is practised on an intensive scale. The colonies are transported in turn to districts where fruit, dandelion, raspberries, sainfoin, white clover and heather abound. The nearby pine forests also offer a rich harvest – a honey more highly priced than that obtained from floral sources. All the colonies are in single-walled hives, the brood chamber holding ten combs of about the same dimensions as the British Standard. At the end of the season, all are brought back and wintered in specially constructed sheds. This arrangement therefore combines the advantages of both worlds – outdoor beekeeping during the active season, and the wintering of colonies in the secure shelter of a beehouse. At Illertissen, at one time, every European race and strain was tested, side by side, for honey production. Finally, one particular strain of Carniolans was retained to the exclusion of all others. An isolation apiary is operated by this firm, in the Bavarian Alps, to retain and ensure the purity of this strain. This concern is undoubtedly the most progressive and successful commercial venture of its kind in Germany. It was, indeed, a revelation to witness what enterprise, unhampered by tradition and prejudice, could achieve in a country where the average colony return is so unbelievably low.

In many respects bee culture in Germany presents an enigma. In so far as apicultural science is concerned, Germany occupied first place up to the outbreak of the last war. This fact was universally acknowledged by scientists throughout the world. In the sphere of practical beekeeping, it lagged behind most civilised countries – judging by the net return obtained. The average annual return per colony is about 9 lb. of surplus honey. I cannot bring myself to believe that the scarcity of nectar-bearing flora is the sole cause that would account for such an insignificant yield. The native bee may to some extent be held partly responsible. The Central European brown bee is actually on the verge of extinction. There are very few strains of the true native left – if any at all. The Carniolan bee has supplanted the bee which was in common use until recently. Sheer economic circumstances and considerations have compelled the adoption of this change; it is a step in the right direction, but only a first step. Bee culture in Germany is unfortunately bound by tradition and prejudice, and it lacks vision and breadth of view. Purely theoretical considerations have completely stifled practical considerations. A glance at a German appliance catalogue will immediately demonstrate the bewildering maze of gadgets and devices – leaving out of consideration the untold number of different sizes of frames and hives – and make evident the futility of the best of efforts at honey production when attempted by such means. Perfection in beekeeping is not found in a multiplicity of appliances, but in simplicity and the elimination of everything not absolutely essential.

My primary task in Germany was devoted to a detailed study of the breeding of bees as carried out in that country. It must be remembered that, outside the German-speaking countries, the subject of breeding improved strains by controlled mating rarely if ever received a mention in any of our journals until quite recently. On the other hand, mating stations have been in operation on the Continent for more than half a century. Dr. U. Kramer initiated *Die Rassenzucht der Schweizer Imker* in 1898. Dr. E. Zander took the movement up in Germany. Leading scientists and beekeepers alike of both these countries have ever since been pre-occupied with the many problems associated with the breeding of bees. Information of great value has necessarily been gathered in the course of 50 years; it is an immense pity that this vast accumulation of experience and information on such a vital subject, is beyond the reach of most beekeepers outside the German-speaking countries. It was clearly an integral part of my task to collect the best information I could. To this end, Dr. Birklein, President of the German *Imkerbund*, kindly rendered me an invaluable service by arranging a visit to the leading Research Institutes, at Erlangen, Frankfurt, Marburg, Celle and Freiberg.

I set out on October 16th on a round of the German Research Institutes. Erlangen, the most renowned of them all, was my first point of call. Erlangen and Zander are synonyms throughout the beekeeping world. Prof. Zander must be regarded as the creator of the Erlangen Research Institute, and the founder in Germany of the movement for breeding an improved native bee. In his view the 'Nigra' embodied the best characteristics a representative German bee should possess. The 'Nigra' varies in colour from jet black to brown. Zander sought his ideal bee in the extreme jet black variant. But many disappointments were awaiting him in this direction. Nevertheless, Erlangen remains to this day one of the few strongholds of the 'Nigra' outside Switzerland.

I next called on H. Gontarski, who is in charge of the Bee Research Institute attached to the University of Frankfurt. Herr Gontarski is one of the most distinguished scientists, and he is well known for his brilliant studies of Nosema. From Oberursel I went on to Marburg. This Research Institute was at that time under the direction of Dr. Dreher, who is also in charge of the central organisation of the *Imkerbund* which controls the breeding of bees. Dr. Dreher possesses a very extensive practical experience and theoretical knowledge of breeding, and his views and articles on this subject command attention. A great number of queens are raised annually by this Institute, to assist beekeepers in securing stock of pure strain.

Celle, situated on the fringe of the Lüneburger Heide, was my next point of call. Dr. E. Wohlgemuth is Director, and Dr. J. Evenius the Senior Research Officer at the Celle Institute. Here again, as at every German Research Station, the utmost importance is attached to the improvement in the bee. To ensure absolute control and purity of strain, the Celle Institute operates a mating station on an island off the German coast in the North Sea.

With the exception of Erlangen, at every Research Station visited in Western Germany the Carniolan bee takes pride of place. However, at Celle three races are maintained, for the purpose of comparative experiments; the 'Nigra' and a strain of Italian origin are kept in addition to the Carniolan. To ensure reliable comparisons, 23 colonies of each race are maintained under identical conditions in one apiary set aside for this purpose. During the season of 1948 and 1949 the averages recorded (which include surplus plus winter stores) expressed in percentages were as follows: Italian, 79.9 per cent; 'Nigra', 85.8 per cent; and Carniolan, 146.1 per cent. The difference in the relative values is very substantial. Only by comparative tests of this kind, carried out over a series of years and on a comprehensive scale, is it possible to determine with certainty the actual value of a race or strain. Indeed, without such tests and a constant re-checking, the breeding of bees is a blindman's game. No real progress is in fact possible.

As already intimated, the native bee of Central Europe is rapidly being displaced by the Carniolan bee. At the Mayen Institute, Dr. Goetze worked on a native strain named 'Hessen'. The origin of this strain seems wrapped in obscurity; some of its characteristics reminded me of the old English native bee. But even Dr. Götze now favours the Carniolan bee. The most noted commercial strains of this bee are known under the following names: Peschetz. Sklenar and Troisek. All three are equally favoured in Germany.

CONCLUSION

After nearly half a century of unremitting efforts, devoted to the improvement of the indigenous bee, a complete change has taken place in favour of the Carniolan bee. The great work inaugurated by Dr. Kramer in 1898 has not by any means lost momentum. But the net results achieved, during the intervening years, have given rise to doubts, hesitation and uncertainty. The *Körsystem*, initiated during the Nazi era, seems to have sounded the death knell of the German native strains. This system of selection (Körung) was based on the assumption that certain characteristics of the bee were an infallible hallmark of its value as a honey producer. Selection according to a pre-determined set of external characteristics, leaving out of account comparative colony tests, was necessarily bound to lead to failure. But I believe it was not the *Körsystem* alone that brought about a deterioration of the native strains in Germany. According to our experience, covering a period of 26 years of controlled mating, the scheme originated by Dr. Kramer was founded on too narrow a basis and on a number of erroneous suppositions.

It is still constantly taught on the Continent that on no account must a queen be used for breeding whose colony has given exceptional yields of surplus, for such queens cannot be relied on to transmit the exceptional honey-producing ability to their offspring. Colonies that produce record yields are termed *Blender* – a colony that blinds by its brilliant performance, but a performance that is not based on heredity, merely on fortuitous circumstances. Therefore the mother of such a colony, when used as a breeder, can lead to nothing but failure and disappointments. In contradistinction, the greatest value is placed on mediocre performance. There is undoubtedly a grain of truth in these assertions; exceptional yields may be purely accidental, or may be the results of a cross-mating of which no visible evidence is shown in the external characteristics of the bees. However, it is one of the axioms in breeding, when dealing with pure stock, that 'like breeds like'. But modern genetics has shown that, in the case of sexually reproduced organisms, there are hardly any instances of absolute uniformity. Some variations must therefore take place, whether we breed from mediocre or exceptional stock. However, by the constant elimination of exceptional performance, no real progress in breeding is possible. To make progress a certainty, it is of paramount importance that a number of queens should be used as breeders annually, for two reasons: there are no means whereby one can ascertain in advance which of a given number of queens, of equally high performance, will in fact prove the best breeder; and secondly, whenever a number of breeders is used, one is enabled to make comparative tests of their progeny, and the actual results obtained will determine the issue. Without constant comparative tests, the breeding of bees is a hopeless gamble.

The same truth holds good with regard to the practice of the use of one drone colony at each mating station, as advocated on the Continent. Here again there is no certainty which colony, or which queen, will furnish the best drones. If an error in the choice is made, the damage caused is irreparable. We therefore keep a minimum of three or four drone colonies at our isolation apiary. The queens heading these drone colonies must necessarily be sisters, selected perhaps from a 100 to 200 colonies. The four sister queens are probably never absolutely identical genetically. Consequently, in the female offspring of these four sets of drones, a wider variation and a wider selection is again secured. In addition, as there are four times the number of drones flying, a more reliable and speedier mating is secured than would otherwise be the case.

The value of mating stations, as operated in Austria, Germany and Switzerland, is at present cast in doubt and uncertainty by many of the foremost authorities on the Continent. After 50 years of ceaseless efforts a deterioration of strains has taken place, instead of an improvement. I believe the suggestions I have put forward will prove a solution to some of the problems that have frustrated years of effort. It may perhaps seem presumption on my part to make these suggestions; however, I have passed through the entire range of difficulties associated with the management of an isolation apiary, and we are able to record progress, though our methods are totally at variance to those used on the Continent.

Mistakes have been made on the Continent in the controlled breeding of bees – at the cost of a heavy price. Nevertheless, a vast fund of practical experience and knowledge, of great value, has been accumulated. The rest of the world may to advantage draw on this knowledge.

1952
ALGERIA – ISRAEL – JORDAN – SYRIA – LEBANON
CYPRUS – GREECE – CRETE
YUGOSLAVIA – SLOVENIA – LIGURIAN ALPS

On February 19th 1952, Br. Adam set out on the journey which was the most strenuous of all not only because of the distances covered and the number of countries visited, but also because from the start constant improvisations had to be made. His first objective was North Africa, but owing to the political tension reigning at that time, he was compelled to restrict his search to Algeria and a few of the oases forming part of that country. At the beginning of April he proceeded by boat to Israel, and thence to Jordan, Syria, Lebanon and Cyprus. Early in June he left Cyprus for Athens and from there explored Attica, the Peloponnese and also Crete. Finally, he made his way from Patras via Arta, Janina, Metzowon in the heart of the Pindus, to Veria, Edessa and to Salonika. After his search of Northern Greece he left for Yugoslavia. By way of Macedonia, Serbia and Croatia he reached Slovenia and the home of the Carniolan bee. In the adjoining Provinces of Carinthia and Styria he collected additional samples before proceeding once more to the Ligurian Alps. On September 20th he returned to England via Le Havre.

NORTH AFRICA

The indigenous honeybee of North Africa is known by a number of names. Naturalists called it *Apis mellifera unicolor* var. *intermissa*. The zoologist, H. v. Buttel-Reepen, gave it the sub-title *intermissa*, for he thought it was an intermediate species between the single-coloured black bee at Madagascar and the variety *lehzeni* of north-west Germany and Scandinavia. Whether this supposition is correct, further research will determine. However, since 1906 this race has been known in scientific literature as *intermissa*.

Frank Benton, of the USA, visited Tunis in 1883 to ascertain the value of the bees found in this part of the world. He collected some queens and called this new variety the 'Tunisian bee', assuming no doubt that this race was confined to Tunisia. John Hewitt visited the same country subsequently and brought the North African bee to the notice of English beekeepers under the name of the 'Punic bee'. In North Africa it is commonly known as the 'Arab bee.'

The distribution of this race in its most typical form is confined to the region of North Africa bounded on the east by the Libyan Desert, on the south by the Sahara, on the west by the Atlantic, and on the north by the Mediterranean. It is therefore isolated on every side by a barrier insuperable to bees. Its native habitat is clearly not limited to Tunisia; it is also indigenous in Tripoli, Algeria and Morocco. However, its main centre of distribution is undoubtedly in the high ground known to the Arab as *tell*; the name 'Tellian bee', first suggested by Ph. J. Baldensberger, would therefore seem to be the most appropriate.

Surprisingly enough reference books have only the scantiest details of the characteristics of the Tellian bee, and the information given is almost all disparaging. In an effort to obtain some first-hand experience of this race, I tried unsuccessfully to import a few queens direct from North Africa over 30 years ago. However, from the information collected in the extreme south of France and Sicily on the 1950 journey, I had high hopes of the Tellian for cross-breeding. My findings in its native habitat confirmed these expectations, which have since been further substantiated by observations made in our own apiaries in 1953. The biometric investigations carried out by Dr. F. Ruttner, on material supplied by him, have corroborated my view on the value of this race for cross-breeding. According to his findings the Tellian incorporates all the known external characteristics of the European races of honeybees.

Ferula thyrsiflora and asphodel by the roadside.

PLATE V

ALGERIA

An arab apiary of ferula hives in the Plain du Sig, Oran.

A classical instance of the extreme nervousness of the Tellian bee.

A ferula hive with a Tellian colony.

53

A pastoral scene on the fringe of the desert.

PLATE VI

SAHARA LAGHOAUT

At Laghouat. A Tellian colony buried in a stack of alfa grass as a protection against the scorching sun.

A ferula hive completely enveloped in a covering of clay, the more common form of protection.

When we set out at the end of February, wintry conditions prevailed almost everywhere. A more violent contrast and transformation than that which I found on stepping ashore at Algiers would be difficult to visualise. The orange blossom was well forward; several eucalyptus were in full bloom – there was in fact a riot of blossom defying description, in gardens and fields, in the woods and primitive bush, and in the hills and the desert. Swarming was in full swing and the main flow at hand.

Professor A. Sturer was at the quayside at Algiers, and also M. Paradeau, one of the most progressive and successful professional beekeepers in North Africa. His preparations during the preceding months, together with his intimate knowledge of local conditions, enabled us to explore Algeria more thoroughly and speedily than would otherwise have been possible. We set to work within a few hours of my arrival.

A series of apiaries was visited in quick succession in every part of Algeria – in the secluded valleys amongst the snow-capped peaks of the Djurjura range, in the primitive bush still found here and there along the Mediterranean seaboard, on the sparsely populated plateau wedged between the Atlas and the Sahara, and on the very fringe of the desert and in the desert itself. We visited a large number of commercial apiaries; these are mainly in the fertile region between the Atlas mountains and the Mediterranean, where the almost boundless citrus groves are found. However, our main search took place in the primitive apiaries in remote parts of the country, where by force of circumstances the Tellian bee has retained its greatest uniformity and purity.

Extensive beekeeping and the use of modern equipment is mainly restricted to the French population, and the progressive commercial apiarists rely on hybrid Italians. The hives are of Langstroth or Dadant pattern. The huge citrus groves (mainly oranges) provide the principal source of nectar. Extraordinary crops are secured in a favourable season and with appropriate management. Considerable yields are obtained also from eucalyptus, rosemary, lavender, thyme and a host of secondary sources. Migratory beekeeping is widely practised by the professional apiarists.

The beekeeping carried out by the natives is of the simplest and most primitive kind imaginable. Throughout the whole of Algeria we never came across any other type of primitive hive than that made of ferula stems. *Ferula thyrsiflora* grows everywhere in profusion, and to gigantic stature. It furnishes the cheapest possible hive material; the mature ferula stems can be had for the gathering in the autumn, and a complete hive costs about 75 francs (about 1s. 9d.). On our journeys we often passed camels and donkeys with loads of these hives on their way to market. In spite of the very primitive mode of beekeeping, the crops secured by these Arab beekeepers probably fall not far short of those obtained in some European countries with modern equipment and by advanced methods. Apart from the possible initial cost of the ferula hives, these Arabs do not incur any expense in producing honey.

In Sicily, where ferula hives are also widely used, some protection from sun and rain is given; the hives are neatly stacked in tiers, four or five on top of each other, perhaps as many as 20 tiers side by side, the whole arrangement forming one huge block of hives. In addition, an open shed provides some protection against extremes of temperature and torrential rain. No such orderly arrangements and elementary safeguards are met in a primitive Arab apiary. Usually the ferula hives are scattered about on the ground with a wanton abandon; often they are disintegrating. Thus exposed to the elements, the bees must thrive or perish. However, they have not only to brave extremes of temperature and torrential rain in winter; they must also defy a host of enemies such as is perhaps not found elsewhere in the world. In the course of ages, in environments of this kind, Nature has relentlessly moulded the Tellian bee as we know it today. But as so often happens, where surpassing qualities are found, these are themselves the direct cause of some serious defects.

With a somewhat subtle unanimity, every work of reference I have seen gives the Tellian bee a deprecatory mention. The general appraisal and recommendation is thus summarised: 'an inferior race in almost every respect, one that should never be imported into any country'. However, more than 70 years have passed by since Frank Benton collected his first queens in Tunisia and, as so often happens, what was at one time discarded as of little value is – with increased knowledge – later deemed to be of supreme importance. Admittedly the Tellian bee is of no value to the amateur beekeeper. But there seems little doubt that it is one of the most valuable races for cross-breeding. Its intrinsic usefulness for this purpose will be largely determined by the care exercised in selecting the breeding stock and – equally important – in the care brought to bear on the crossing, in order to bring out the best qualities of the race.

The pure Tellian bee is black – jet black – and if anything more so than the 'Nigra' of Swiss origin; its blackness is accentuated by the scanty tomenta and over-hair. It is perhaps slightly larger than its nearest cousin, *Apis mellifera* var. *sicula*. They are jet black, long and slender and very pointed – quite unlike the plump Italian or ponderous Carniolan queens in shape. Both queens and bees are quick in movement and liable to extreme nervousness when manipulated. Indeed, when a hive is opened, the bees are disposed to 'boil over' and 'mill around' inside the brood chamber in a most alarming manner. But if left a few minutes and given a chance to calm down, they will thereafter submit to manipulation as readily as any of the common bees of Northern Europe. They can be bad tempered, but not more so than the black bees of Southern France which used to be imported in such large numbers into this country. Though we came across some extremely bad tempered Tellians on our search, we discovered at the same time a few strains which could be handled with the greatest impunity. In my estimation the most serious defects of the Tellians are: (1) extreme swarming tendency, (2) a highly developed susceptibility to brood diseases, (3) a lavish use of propolis, (4) watery cappings. Against these defects must be set unparalleled stamina, fertility and foraging power.

The extreme addiction to swarming of the Tellian is doubtless a direct effect of its amazing stamina and fertility. The pronounced innate susceptibility to brood diseases is a defect of nearly every variety of the common European dark bee, particularly the French ones. This defect is, however, more marked in the Tellian than in the French bee. There are in fact a great many close similarities between these two races – for instance the lavish use of propolis. In every characteristic (except cappings) a close relationship can be traced, but the qualities are more pronounced in the Tellian.

The fecundity of the Tellian is remarkable. But extreme fertility is of no avail unless it is coupled with a high degree of stamina, and it is in this very quality that the Tellian surpasses every other race. Moreover, stamina is the source of a whole series of desirable traits, longevity, hardiness, wing-power, etc. Observations made in 1953 lead me to believe that the Tellian is the longest-lived bee. I also noted that it is active at temperatures at which no other honeybees would dare to venture forth, not even Carniolans.

As already indicated, the Tellian has not only to brave extremes of climatic conditions in its native habitat, but it must also withstand the ravages of innumerable enemies. The huge jet black pollen beetle, *Cetonia opaca*, unknown in Northern Europe, is an ever-present menace, and will, if it can find its way into a hive, wreak sad havoc among the combs. The bees seem fairly helpless in face of this venture. They are equally defenceless against the voracious blue-cheeked bee-eater, *Merops superciliosus* – one of the most lovely birds in creation, but a deadly enemy of the honeybee. This bird thrives on bees, though it will occasionally include a wasp or two in its diet. The loss of bees is all the greater because *Merops superciliosus* does not live singly, but in flocks of up to a 100 birds. It is estimated that a flock of this size will dispose of a pound of bees in a day. The bee-eater is a seasonal menace, for it migrates in September to the Cape of Good Hope and re-appears in March. The Oriental hornet is represented in full force in North Africa; the blind ant *(Dorylus fulvus)* must, however, be regarded as the most treacherous enemy. This insect will make its way into a hive unnoticed by gnawing a hole through the bottom board, and before the beekeeper is aware that something is amiss, the colony has perished and the invader has made good his escape. Lizards and toads are constantly around the hives. When lifting the roofs off a hive, it is not uncommon to find a batch of lizards scampering away. Wax moths are a serious problem in every sub-tropical country; a colony which is not resistant, and which cannot maintain its strength through the summer months, has little chance of escaping destruction from their ravages.

It is often claimed that the production of parthenogenetic or impaternate females is a common phenomena in Tellian colonies. I have not so far found any evidence to support this view.

My search in Algeria would not have been complete without exploring some of the oases in the Sahara, and I should have missed one of the best opportunities found in Nature to study the effects of many centuries of inbreeding on the honeybee. Moreover, there was every likelihood that, in the complete isolation and added rigours of an oasis, a strain of the type required for cross-breeding would be found. Though my time was drawing to a close, I nevertheless decided to visit Laghouat, Ghardaia, Bou-Saada and perhaps some less well known oases *en route* if at all possible.

Since my arrival in North Africa I had seen much of the wonderful flora of Algeria: pinky white drifts of asphodel; wide expanses carpeted in bright orange by the native marigold, *Calendula algeriensis: Oxalis corniculata rubra* and *variabilis*, in great masses; giant clumps of the glistening white *Erica arborea*; and thymes in mauve, blue and purple. Perhaps the sections of primitive bush along the Mediterranean seaboard contain the most fascinating collection of wild flower and shurbs within any given space. The most important nectar-bearing sources of this sub-tropical jungle are rosemary and lavender, *Lavendula stacchus*, which thrive here in a profusion hardly seen elsewhere. But on our way south into the Sahara we found a totally different kind of wild flora: the desert in bloom, in its full but ephemeral springtime glory – a dense carpet of desert flowers, stretching to the horizon in every direction. The air was heavily laden with the sweet scent of honey, and the traffic of insects gave the impression of a large number of swarms crossing to and fro overhead. But there were no honeybees amongst this busy throng. In these desolate regions they could not survive after the brief, brilliant spell of spring.

At Laghouat I found about 50 colonies of bees, owned by three beekeepers: one a Christian, another a Jew and the third a Mohammedan. At the apiary owned by the Christian, the bees were in modern hives and kept with a meticulous and finicky solicitude characteristic of an amateur. At the apiary belonging to the Hebrew, I found a conglomeration of different hives, as well as boxes of every size and shape suspended upside-down amongst the branches of tangerine trees; these contained newly-hived swarms. Dead virgin queens could be picked up by the dozen under these boxes. The third beekeeper, a retired Arab officer of the French colonial forces, graciously allowed us to view the seclusion of his bee garden, but not until the customary formalities had been duly observed. His apiary consisted of ferula hives, of traditional shape and size, except that for some reason they were encased in a heavy coating of clay. The old Arab proudly pointed to one hive, hidden in a mountain of alfa grass, which furnished no less than seven swarms the previous year. At the end of the swarming season no more than 200 or 300 bees were left. Yet this miniature colony survived and filled the hive with new comb, brood and honey – ready to respond again to the impulse of colonisation. Inbreeding – perhaps since time immemorial – had in this instance no harmful effect on viability of the brood and on the stamina of the bees. Indeed, it was at Laghouat that we found the most powerful stocks of pure Tellians, covering 20 combs of Dadant size in March. The bees at this oasis were remarkably good-tempered, notwithstanding the fact that at the time of my visit a fierce sandstorm was raging.

Owing to the violence of the storm there was no possible chance of penetrating deeper into the Sahara. I had to retrace my steps, and even the journey north, to Bou-Saada, proved a perilous venture. The extreme heat, accompanied by a following sirocco, further accentuated by the difficulties of the desert track part of the way, proved almost my undoing, as there was no water within miles to replenish losses from the car radiator. Though I endured extremes of heat and hazards of one kind or another during the subsequent months, the ordeal of the trip from Laghouat to Bou-Saada was never equalled. I reached Algiers on March 30th, and next morning left for Marseilles, to re-embark on April 2nd for Israel.

I have refrained from a more detailed description of the less obvious characteristics of the Tellian bee, for my investigations are not yet concluded. However, all the findings I have made up to now indicate that the Tellian is a primary race, and that the numerous varieties of brown or black bees – at least those of Western Europe – have in the course of time evolved from the Tellian.

ISRAEL

After a rather unpleasant seven days at sea, Palestine – the land flowing with milk and honey – was reached on April 8th. I spent the night on Mount Carmel, and on the journey to Tel Aviv next morning, the Holy Land revealed itself in all its springtime glory. I was told that the extraordinary profusion of wild flower which I saw had not been known for nearly half a century; it was due to an exceptionally heavy rainfall the previous winter.

The route to Tel Aviv took me through the most fertile part of Israel, through the Plain of Saron extending southwards from Mount Carmel. A belt of orange groves, about 20 miles wide, stretches all the way to Jaffa and beyond. The groves were in full bloom, and the heavy fragrance of orange blossom pervaded the countryside. I was told that the nectar flow had almost reached its greatest intensity, and that beekeepers were already busy extracting.

At the Ministry of Agriculture in Tel Aviv I was introduced to Mr. D. Ardi, Apicultural Adviser to the Government. Plans were quickly drawn up for the search throughout Israel, and it was arranged that Mr. D. Ardi should act as my guide. I wish to record here my grateful appreciation to him, for his help and hospitality.

The dynamic drive of this newly-formed State was in evidence everywhere. Economic problems are being solved in the most direct and effective way possible. Perhaps the most notable example is the action taken by the Israeli Ministry of Agriculture to supply the highest quality breeding stock to beekeepers throughout the country. The breeding stock is raised at Government-owned mating stations, the most important being at Hefzebah, near the site of ancient Caesarea. By law no other bee may be kept within three miles of this mating station. Breeding stock of a specially selected strain of Italians is sent out from Hafzebah; this strain was exhaustively tested over a period of years in the climatic conditions of Israel, side by side with many strains from various sources, before it was generally adopted. By this course of action the Israeli Government is assisting the craft in the most effective way possible.

It is occasionally claimed that Israel possesses its own indigenous race of bees, but more comprehensive enquiries showed that there is no clear-cut difference between the bees found in Lebanon, Syria and Palestine. The slight variations do not warrant a special classification. Geographically, Israel is part of Syria, and there are no natural barriers which would prevent an intermingling if there had been more than one indigenous race.

The Syrian bee, *Apis mellifera* var. *syriaca*, closely resembles the Cyprian; the two races are, however, quite distinct, although closely related. The Syrian bee is smaller, and it shows every defect of the Cyprian in an intensified form – particularly temper. In my estimation the temper of the Syrian deprives this race of any practical value it might otherwise possess, although – unlike some European races – it will not attack unless interfered with. Primitive beekeeping is therefore well able to get along with this bee, for beyond the annual taking of the honey at the end of the season (when colony strength is at its minimum) no interference is called for. But the manipulations demanded by modern beekeeping do not seem feasible with Syrian colonies. Even miniature colonies covering only a few combs will not tolerate disturbance, as I found by experience. Moreover, a swarm of angry bees will pursue and attack any living creature within reach. This habit of attacking *en masse* at great distances from the hive is a very dangerous trait. Tellians, Cyprians and some French strains also show it, but to a much smaller degree.

The pure Syrian is an elegant bee. The abdomen is very pointed, and the first three dorsal segments are a clear lemon yellow. Tomenta and over-hair have a silvery sheen, and the scutellum is bright lemon.

The fecundity of Syrian queens is prodigious – too much so. The bees are good foragers and have great stamina. They are, however, given to excessive swarming, and when the swarming impulse has taken hold of a colony, it will construct an enormous number of queen cells, often hundreds of them. One of the Syrian's most noted good qualities is its intrepid defence of its home.

The true Syrian is distinct in appearance and biological characteristics from all other races. It is, however, no longer easy to find colonies of the pure Syrian. In Israel itself they can perhaps only be found in Upper Galilee, in the region between Lake Hula and Metulla. In the Jordan sector they are more common. But in Northern Lebanon and Syria the influence of the Anatolian bee can be clearly discerned. In fact there is considerable variation even in colonies immediately north of Beirut. Hybrids predominate everywhere in Israel, for strenuous efforts are being made to supplant the indigenous bee.

There are few Israeli beekeepers who regard the introduction of Italians as a serious mistake. The well worn arguments in support of the indigenous bee are brought forward in Israel, as in many other countries. We visited one of the adherents of the Syrian bee, and were given a demonstration of their docility. I left unconvinced. In my estimation the Syrian bee has not one redeeming quality which would atone for its irascibility. Though I was often assured that really docile strains do exist, I never came across any on my search. On entering an apiary where Syrian bees were kept in modern hives, one was instantly confronted by a horde of angry, hissing bees, and a throng of them would pursue one for a considerable distance after leaving the apiary. This extreme viciousness is sometimes regarded as eminently desirable: one of the most able Arab beekeepers assured me that he only got a honey crop because the temper of his bees prevented unwarranted persons interfering with his hives.

In 1952 Israel possessed about 33,000 colonies of bees, and efforts are in progress to double this number within a few years. The required material is being imported from America. Langstroth equipment is used exclusively, and to ensure economy and simplification in management, full-depth brood chambers have to serve as supers. Primitive hives are only found in isolated Arab villages.

Commercial beekeeping is mainly confined to the communal co-operative settlements or kibbutzim. Some of the kibbutz operate up to a 1,000 colonies. Emphasis is placed on intensive rather than extensive beekeeping; the scarcity of timber, the high cost of imported hives and general economic conditions preclude any haphazard keeping of bees. The main honey crop is from the orange blossom, which yields 20-30 kg. per colony. At the end of April, or early in May, the hives are taken from the orange groves in the coastal plain to the hills and mountains of Galilee, to gather the second crop from the wild flowers, the most important being acacia, cactus, lavender, wild carrot, sage, thyme and a great variety of thistles. The second crop averages a further 20-30 kg. per colony. Commercial beekeeping undoubtedly has a promising future in Palestine.

As one would expect, the honey crop in the Levantine countries depends largely on the rainfall during the brief winter months. This is true for the orange blossom, and even more so for the crop from the wild flowers. Yet hopes raised by an abundance of rain may in the end be dashed to the ground by the dreaded khamseen at blossoming time. This happened in 1952. All the Middle East countries had an exceptionally heavy rainfall the previous winter, and the orange groves were laden with an exceptional abundance of blossom. But as nectar secretion reached its maximum intensity, the hot khamseen from the desert shrivelled the blossom in a few hours. Instead of a record crop, only 6 kg. per hive was secured – the lowest average for ten years. However, the wild flowers on the hills and mountains were unaffected, and an exceptional crop was secured from them.

From mid-July until November, when the rainy season starts, there is no nectar or pollen; during this period the colonies must also fight for survival against hornets and wax moths. This fight is a grim one: the colonies are first weakened by the hornets, and the wax moths give the coup de grâce. In spite of every effort by the beekeepers to combat the hornets, by poison baits and the destruction of nests, the annual loss of colonies is about 10 per cent – in some seasons even 30 per cent. Some beekeepers have been compelled to move entire apiaries to areas less heavily infested with hornets.

The rain and cold in November bring to an end the fierce struggle between the honeybee and its enemies and, with the beginning of the rainy season, a new lease of life sets in for the bee. In the maritime regions the carob *(Ceratonia siliqua)* and the loquat *(Eriobothrya japonica)* yield abundant nectar and pollen when the weather is favourable. In the higher regions severe though brief conditions are not uncommon; winter, however, offers no serious problem to the beekeeper.

I had heard so much in years gone by of the Syrian bee through the kindness of Fr. Maurus Massé who, during his sojourn at our Monastery at Abou-Gosch, tried to make the best of this race. He had little success, and small reward for his efforts, and I am now no longer surprised at his failure.

JORDAN

On April 19th I crossed over to Jordan, to our Monastery of St. Benoît on Mount Olivet. This is south-west of Jerusalem, and gave a perfect view of the Old City and the Temple Area. Until quite recently Syrian bees were kept at the Monastery, in modern hives, but with no great success.

The Arabs have great faith in their native bee. Over and over again I was assured that there were two distinct varieties of indigenous bees, one of which builds combs in the shape of the moon, and the other in the form of a furrow. It was further claimed that the former was of good temper, but short-lived and a poor forager. The second kind was of vile temper, but long-lived and a great honey gatherer. Unfortunately, this ready differentiation will not bear close scrutiny. Expressed without the Oriental simile, a cast hived in a clay cylinder will build comb parallel to the entrance, and therefore in the shape of a more or less perfect circle. On the other hand, a prime swarm will at once occupy the greater part of the cylinder and build comb at right-angles to the entrance – or cold-way in the more prosaic language of the European. A cast has little chance of escaping the ravages of hornet and wax moth, and is therefore in the eyes of the uninitiated short-lived and not very valuable as a honey gatherer. This notion of there being two distinct varieties of bees, of one and the same indigenous race, is surprisingly widespread in the Middle East. The same view, based on the same differentiation, is held in Cyprus.

ISRAEL – Earthenwear hives near the village of Abu Gosh; They are also used in Jordan and Lebanon.

By the Lake of Genezareth.

PLATE VII

ISRAEL
JORDAN
LEBANON

Jordan: An unusual type of shelter near Bethlehem. The sun-baked clay hives, as commonly used by the Arabs, are strongly made and very roomy.

Lebanon: Tubular wickerwork hives are in common use here.

An Arab hive of sun-baked clay.

Considerable efforts have been made in recent years to introduce the modern hive into Jordan. But without introducing a more manageable bee at the same time, these well meant endeavours seem doomed to failure. There is nothing to be gained by putting Syrian bees into a modern hive and then – because of their unmanageability – leaving them to their own devices. They might as well be hived in a clay cylinder. The net return in surplus honey would show no material difference, but there would be a substantial difference in the cost of production between the modern and primitive way of keeping bees. In a country without timber, a sustained effort to introduce a bee more suitable to modern methods of management will probably never be made, the cost of a frame hive will never be justified. The sun-baked clay cylinders cost next to nothing and, if large enough, they offer a satisfactory home for the Syrian bee.

My enquiries in Jordan took me to a great many primitive apiaries, but I came across none containing many colonies; there were a dozen at most, but more often only two to four. The clay hives are substantially constructed and of no mean capacity, and thus well suited to the extremes of temperature and the ability of the native bee. They are 26 in. long and 12 in. in diameter internally. The walls are a full 2 in. thick. Less common are the hives of stoneware, made in the shape of an Oriental water jar of about 2 gallons capacity. The narrow neck forms the entrance. The jars rest on their sides, and the opening for removing the honey is at the back which is fitted with a detachable disc. These stoneware hives have the advantage of great durability, and also provide an almost complete safeguard against the many troublesome pests. But hives of stoneware require shelter from the direct rays of the sun, whereas the clay cylinders do not. These stoneware hives seem to be confined to Jordan and Lebanon; at least I did not see them anywhere else.

On May 7th I left Jerusalem for Syria and Lebanon, via Jericho and Amman. As I approached Jericho, the wheat harvest was already in full swing. The season was advancing rapidly. The lilies of the fields had gone until the next return of spring, and the landscape was brown and seared. But on leaving Israel I was again confronted with some of the loveliest scenery imaginable in the verdant valley of the Wadi Salt, along which the road winds its way to Amman after leaving the Plain of Jericho. This narrow valley, set amidst the desolate hills of ancient Moab, with its profusion of wild flowers, its masses of oleander in full bloom, and the vivid scarlet waxy blossom of the pomegranate everywhere, combined to form a picture of unforgettable loveliness. In this beautiful setting, the Jordan Department of Agriculture recently established an experimental apiary between Suveille and En Salt.

When I arrived in Amman I paid the expected call at the Department of Agriculture, and then set out of the hazardous trek across the desert to Damascus.

SYRIA AND LEBANON

By the time I entered Syria I had gathered quite a valuable collection of samples for the Bee Department at Rothamsted – of value for biometric studies, but for no other purpose. However, the Syrian Customs thought otherwise. The many cases full of glass tubes, each with its preservative, label and number, seemed to them too valuable to pass without payment of a heavy deposit. And I was on the way to Damascus, where they held that such things could be sold. After two hours' delay, spent in the insufferable heat of the Arabian Desert, I was allowed to proceed (having paid substantially for the trouble I had caused), with every case securely fastened with a lead seal. This was but the beginning of the difficulties these samples involved, until more enlightened Customs were reached months later.

Among the marvellous vegetation of Lebanon must be counted many wild clovers. I had already seen many varieties new to me when in Galilee, but they grow in much greater profusion in Lebanon. Indeed, I was told in Beirut that no record of all the species had yet been made; it is thought that there may be 150 or more. My attention was attracted particularly to two miniature species, one white and one red. Neither grows more than 3 in. high, but the profusion of blossom is amazing; the clover-heads form dense carpets of white or purple. When I first crossed the highest part of the Lebanon mountains coming from Damascus, huge patches of purple caught my eye, which proved to be the miniature red clover in full bloom; its value as a source of nectar was instantly apparent, for it was alive with bees. Indeed, I had never before witnessed so many honeybees foraging with such intensity in a specified area. Moreover, they must have come from a great distance, for on this otherwise bare and bleak mountain plateau no hives could be seen for miles. The miniature white clover is just as valuable as a nectar source. Both species thrive at sea level and at higher altitudes, but the tiny red clover seem to do best at about 3,000 feet, and on the poor soil found on the Lebanon mountains. The white species (but not the red) I observed in Cyprus at the higher altitude of Troodos.

The flora of Lebanon is more luxuriant, and if anything more varied, than that of Israel. The mountainous country ensures a heavier rainfall, and the high humidity and the oppressive steamy heat impart to the low-lying maritime regions a genuine tropical character throughout the summer. The belt of citrus groves, banana and loquat plantations along the seashore furnish one of the main sources of nectar, but the extremely varied nectar-yielding flora of hill and mountainside provide a honey harvest no less rich. Indeed, I believe that Lebanon has at its command one of the richest and most varied bee flora in the world.

The potentialities of beekeeping in Lebanon are reflected in the size of the primitive hives. Tradition and experience over the centuries have doubtless demonstrated the advantages of a hive which will hold a honey yield much above the average secured in other countries. The Lebanese hives are tubular, and measure a full 4 ft. in length and 11 in. in diameter. They are not made of timber, clay or stoneware, or of ferula stems, as in the other countries I had visited, but of wicker with a thin finishing coat of clay. Stiff wooden members are woven into the wicker-work lengthwise, to give the tubular construction the necessary stability and rigidity. These wicker hives cannot be stood directly on the ground (particularly in a humid climate); they are placed individually on shelves, a series of shelves being built one above the other, in an open shed with some sort of roof. At Baalbeck – renowned for its honey, as well as for its unique temple ruins – I saw the most capacious primitive hives of all; they were made of wood, and were no less than 5 ft. in length and 1 ft. in height and width internally.

Modern hives (Langstroth and Dadant) are in fairly common use throughout Lebanon. The Government is doing everything possible to encourage a still wider adoption of modern equipment and advanced methods of bee culture.

The native bee leaves much to be desired. Though it is not quite so irascible as the bee found in Israel, it resents interference. There is a marked difference in colour, size, temper and general behaviour of the Syrian bee north of Beirut. There have been some imports, but I am inclined to ascribe these variations to the influence of the Anatolian bee. Something useful might perhaps be evolved from this heterogeneous collection by selective breeding, but it is questionable whether the labour entailed would be justified. A good reliable strain of *ligustica*, and a distribution of breeding stock on the lines carried out in the adjoining country to the south, would seem to be the right solution. Such a course would yield quick and reliable results, with a minimum of expenditure.

Lebanon is a land of incomparable scenery, and it would be hard to find another of equal size with such a varied climate and such a rich flora. It is a country where bee culture should flourish as nowhere else in the Middle East.

CYPRUS

It was with keen expectations that I visited Cyprus. More than 33 years had passed since the first consignment of Cyprian queens reached Buckfast, and a number were imported later. I was therefore fairly well versed in the idiosyncracies of this race *(Apis mellifera* var. *cypria)*, but there were several important problems which could only be solved by studying it in its native habitat. Moreover, there was good cause to suspect that a thorough search would reveal isolated strains of a more benign disposition than any we had so far possessed.

Cyprus was reached on May 17th. Representatives of the Department of Agriculture kindly offered me every assistance when I disembarked at Limassol. However, nothing useful could be done that day, for I had hardly arrived before it began to rain, and it rained with tropical intensity. This downpour was not only unseasonable but also most inopportune, as the corn was still being harvested. It was, however, a welcome change to me after the steamy heat of Beirut.

I returned to Nicosia on the following Monday to call on the Department of Agriculture. Immediately on my arrival, the Department kindly gave me a list of all the important apiaries in the island, complete with the number of colonies in each and the type of hive. After a brief consultation, Mr. Osman Nouro drew up an itinerary and issued instructions to the District Officers concerned. The first week was taken up by exploring the northern and central sections, and the search was then extended to the districts of Famagusta, Larnaca, Limassol, Paphos and Lefka. On June 4th I left for Greece from Larnaca. Thanks to the efficient arrangements and the willing co-operation of the various District Officers, I was able to carry out the search not only expeditiously but very thoroughly, and Mr. S. A. L. Thompson also made a substantial contribution to the success of my efforts. I shall also always recall with pleasure the visits to his mountain chalet above Kyrenia, and the marvellous view from his home of Cilicia and the snow-capped peaks of the Taurus in the far distance.

The nectar-yielding flora of Cyprus is fairly varied, but it cannot be compared with that of Lebanon. Moisture is lacking, and there are no permanent rivers. The central Plain – the Messaoria – offers only a bare subsistence to bees for most of the year; it is barren and seared from the end of May until the rains return. The hills and valleys, and the two mountain ranges which extend in parallel lines from east to west of the Plain, offer a much richer provender. The highest peak of the Troodos range to the south reaches 6,406 feet; the Kyrenia range to the north is lower.

The main honey crop is derived from fruit blossom, citrus, thistles and the wild thyme. Owing to the lack of moisture, the clover is useless to the bees, and it is probably for the same reason that the carob (*Ceratonia siliqua*) which is much prized as a nectar source in Sicily, does not yield freely here. This is most unfortunate, for Cyprus is famed for its carob trees; there are about two million of them and, unlike most trees, they seem to thrive everywhere. There are many secondary sources of nectar, from the commencement of the winter rains until the seasonal drought. Bees can collect enough to meet their needs throughout the winter – from the loquat, *Acacia* and *Eucalyptus* which yield in December, then from the various species of dandelion, bean and *Anchusa*, and towards the spring from *Oxalis*, rosemary, sage, etc.

The extensive citrus groves are centred near Famagusta, Limassol and Lefka. The wild thyme, *Thymus capitatus*, the same species from which the famous Hymettus honey is derived, thrives only on bare and parched hillsides, where nothing of much value could subsist. The many species of thistle are mainly found in the more arid sections of the country. Some of them are lovely; the most beautiful of all, found everywhere by the roadside at the end of May, is clothed in a heavenly blue – the slender stem, leaves and all.

Nature has not been particularly indulgent to the honeybee in Cyprus. Except among the orange groves there are no heavy nectar flows. The native bee, by dint of effort, is able to make a living during the greater part of the year, but the amount of surplus gathered is small.

There are about 22,000 colonies of bees in Cyprus, less than 2,000 of them in modern hives. Efforts are in progress to further modern methods of bee culture, and regular courses are given on advanced beekeeping at the Central Experimental Farm in Morphou. There is a small plant for manufacturing comb foundation at this Farm – the only source of it in the island. Apiaries with modern equipment are largely owned by the great fruit-growing concerns. The beekeeping and queen-rearing establishment belonging to Mr. S. A. L. Thompson, at Jingen Bahchesi, Kyrenia, is probably the most progressive of its kind.

The primitive hives in Cyprus are of either burnt or sun-baked clay; they are tubular, about 30 x 10 in. internally. Apiaries containing 100-150 colonies are quite common; the clay tubes are stacked and joined into one solid block, like the individual bricks in a wall. They are usually tiered four or five high, and a large collection of them often resembles a long boundary wall; the roofing tiles which are usually placed on top help to complete the illusion. Small apiaries are uncommon in Cyprus. In some villages, for instance, Paphos, one may occasionally find hives built into the walls of houses, the hives opening on the inside into a bedroom or living room. Though *Ferula thyrsifolia* thrives in Cyprus, it is not used for hives; the more durable clay is preferred.

It is not known when or whence the first colony of bees was brought to Cyprus. The possibility of a vagrant swarm flying from the mainland must be excluded, since Asia Minor is 40 miles and Syria 60 miles away. There is some evidence indicating a descent from Egyptian stock; Cyprus was first occupied by the Egyptians in 1450 BC, and it is known that about 850 years later there were bees on the Island, because Herodotus refers to a swarm which had taken possession of a skull suspended before the temple of Aphrodite. The attention of modern apiculture was first drawn to the Cyprian bee in 1866.

The Cyprian bee is midway in size between the Italian and Syrian. The colour of the first three dorsal segments is a clear bright orange; the fourth and fifth segments are also orange, but only near the ventral plates. Each of the first three dorsal segments has a sharply defined black rim, which is narrowest on the first and widest on the third segment. The colour of the three posterior dorsal segments is a decided black, which tends to enhance the orange of the first three segments. The ventral plates (except the posterior two) are usually a transparent orange without any trace of a darker colouration: this is one of the most characteristic markings of the Cyprian. The scutellum is pale orange, and the over-hair and tomenta are buff.

The queens are considerably smaller than any of European origin. Their colour and markings are much more constant, and the markings so definite, that a Cyprian queen can readily be identified. The abdomen is pale orange, but each dorsal segment bears a narrow, sharply defined crescent-shaped black rim. A somewhat similar marking is occasionally observed in a common hybrid queen, but the bands are then wider and not so sharply defined. Although they are small, Cyprian queens are exceedingly prolific. Their fecundity only reaches its maximum, however, when they are crossed with another race.

Contrary to expectations, pure Cyprians are not adicted to swarming. This would be fatal in their native home. Under the swarming impulse they usually construct a great number of queen cells – often several hundred – and they tend to build them in clusters resembling a miniature bunch of grapes. The breeding power of this race is truly prodigious, and more honey is devoted to brood rearing than pleases the beekeeper, but this must be regarded as a device of Nature to ensure the survival of individual colonies in their native habitat. Cyprians are hardy, long-lived and endowed with great foraging abilities. Their cappings of honey are dark and watery in appearance. They construct little or no brace-comb; they are disposed to use propolis freely, but fortunately not usually the sticky resinous kind, but a compound of propolis, cappings, etc., which does not readily adhere to one's fingers. Lumps of this mixture are often deposited along the entrance in the autumn. Cyprians pass through the winter more safely than any other race, even in our northern climate (although their native home is in the sub-tropics); this is one of their outstanding characteristics. I have never known a Cyprian colony, pure or first-cross, fail to come through the winter.

Perhaps nothing had made the Cyprian bee more unpopular than its irritability. Most strains strongly resent any interference, and this irascibility is just as pronounced in its native habitat. Records of the first imports into Europe, however, laid stress on its remarkable *docility*, and I found that there are still such good-tempered strains in the Island.

Although the Cyprian is probably the most homozygous race known, my enquiry has revealed a measure of variation. There are many deep valleys where individual isolation is as complete as that of the Island itself. These isolated pockets hold the material for the further improvement of the Cyprian race; it should be possible by suitable selection to develop strains as gentle and as tolerant of manipulation as any Italian.

The absolute isolation and the harsh environment of the Island have together given us a priceless asset, and to the enterprising geneticist Cyprus is a veritable Treasure Island. However, the thousands of years of inbreeding between relatively few colonies have in a measure masked the potentialities of the race, and experience leads me to believe that the pent-up qualities of the Cyprian will only unfold to the full in cross-breeding. But I must emphasise that although they are of incomparable value in developing new strains, pure Cyprians are useless to the average beekeeper.

Beekeeping in Cyprus is favoured by one unique blessing – the complete absence of disease. To maintain this good fortune, and to ensure the continued purity of the Cyprian race, imports of queens and bees are strictly prohibited.

GREECE

After two days at sea, we sighted Cape Sunium about noon on June 6th. Athens was reached in the late afternoon, and what proved to be the most exacting and strenuous three weeks of my search lay immediately ahead.

Beyond the bare information that there are more colonies relative to the population (about one for every ten inhabitants) in Greece than in any other country, little was known of beekeeping conditions in this extreme section of south-eastern Europe. But the large number of colonies indicated a certain measure of apicultural prosperity, although not necessarily a substantial surplus yield per colony, which would pre-suppose amongst other things an indigenous bee of outstanding abilities. I was not left in doubt on this point for long.

The day after my arrival found us exploring Attica, as far south as Cape Sunium, with Dr. A. Typaldos-Xydias and Mr. C. Michaelides. Dr. Xydias, who met me the day before at the Piraeus, has been for many years Technical Adviser to the Ministry of Agriculture and may be regarded as the father of modern apiculture in Greece; indeed, I realised daily during the next few weeks that Dr. Xydias is known and revered by every Greek beekeeper.

Not a boundary wall, but a stack of about 150 cylindrical hives south of Nicosia.

In Crete the basket hive with movable combs is used even more widely than on the mainland. A hive of identical size and shape made of burnt clay is also in use here.

CYPRUS

PLATE VIII

CRETE

The classical basket hive of Greece, probably the oldest beehive with movable combs. The earthenware type of smoker may well be as ancient.

An idyllic scene – an ancient bee garden with Mount Hymettus in the background – The kind Solon saw about 600 BC.

PLATE IX

GREECE

A small section of an apiary near Philippi with more than 400 colonies brought from Thasos.

A migratory apiary north-east of Salonica. The seasonal transportation of the bees to a series of crops plays an important role in Greek beekeeping.

77

Our journeys took us twice to the Peloponnesus, and then on the last visit from Patras to Missolonghi, Arta, Janina and Konitse; thereafter to Metsovon in the heart of the Pindus range, and on to Kalambaka, Grevena, Kozania, Veria, Edessa, Salonica and the section of country north-east of that city. The trip to Crete I made alone, as the Agricultural Officials of the island furnished all the assistance required. Arrangements were already made for a visit to a few of the islands in the Aegean Sea, to which both Dr. Xydias and I attached great importance, since it is here – as in Cyprus – that the most valuable breeding stock is likely to be found. Unfortunately, in the end I had no time for this visit.

The ancient Athenians, we are told, were constantly praising four things: their honey, their figs, their myrtle berries and the Propylaea. The honey the Athenians were so proud of was gathered on Mount Hymettus, immediately east of the city. It is derived from the mountain thyme, *Thymus capitatus* and is highly aromatic, with a heavy body and a light amber colour: a most delicious honey indeed, but not one which will always appeal to a palate used to the evanescent flavours of our paler northern honeys. Wild thyme is not confined to Mount Hymettus; it is common throughout Southern Greece, the Peloponnesus and Crete, where it is the principal nectar source. In these regions it thrives on any bare, rocky and otherwise barren hillside, where nothing else can subsist for lack of moisture. At the time of my arrival it had just commenced to bloom, and at some of the apiaries I visited the air was laden with the rich scent of the newly-gathered nectar. However, I was told that it was not secreting heavily for lack of the necessary atmospheric humidity.

Groves of orange and lemon abound in the maritime regions of Southern Greece but, except near Arta, none are as large as those in the Middle East and North Africa. Other varieties of fruit of value to the bees are confined to the northern part of the country; there are extensive plantations between Veria and Naoussa. It is indeed in the north of Greece that the heaviest crops of honey are secured. The main sources are clover, sweet chestnut, wild sage, mountain savory and honeydew. Crete has an extremely abundant and varied nectar-bearing flora, with many species of *Erica*; these seemed to be absent in the Levant.

Greece possesses approximately 700,000 colonies of bees, and I was greatly impressed by the high standard of efficiency of its bee culture – the modern (with Langstroth) as well as the primitive. In Northern Europe beekeeping is usually regarded as a sideline, or as a pleasant hobby, and beekeepers often have only three or four hives. Not so in Greece! There are probably more professional beekeepers in Macedonia than anywhere else in Europe. Migratory beekeeping is the accepted thing, and it is practised on a grand scale with most laudable results. I was told that averages of 100 kg. are not uncommon. From a good vantage point some 30 miles north-east of Salonica it was possible to pick out apiaries containing altogether no less than 2,000 colonies – the area was literally teeming with bees. To the west, beyond Edessa, in well nigh inaccessible regions adjoining Albania, extensive apiaries were tucked away in the folds of the hills everywhere, and the thousands of colonies in them had just been brought there from long distances. Now and again one could see equally large apiaries of primitive hives, which had also been brought to these inhospitable regions. Professional apiarists, modern and primitive alike, rely on migratory beekeeping for a dependable income.

The primitive beekeeping in Greece is instructive, and historically of great interest. We know that the basket hive of today was in common use in Greece more than 3,000 years ago, and that the principle of the movable comb, re-discovered about a hundred years ago, was in fact employed in this hive by the ancient Greeks. The hive is constructed of wickerwork, and has the same shape as an earthenware flower pot. It is 23 in. deep, 5 in. across at the top and 12 in. at the bottom (internally). Nine bars – $1\frac{1}{2}$ in. wide to give the correct spacing – fit across the brim. The combs are attached to these bars, exactly as in the hive invented by Dzierzon about the middle of the last century. With a little extra care, each of the nine combs can be examined individually as freely as the combs of a modern frame hive. Moreover, the shape of this Greek hive corresponds more closely than any modern rectangular one to the natural inclinations of the bees. In Greece the baskets are given a fairly substantial external and internal coating of clay, whereas in Crete – for some reason I was never able to discover – a thin coating only is applied, internally and for about 2 in.

around the bottom externally. In Crete one occasionally sees earthenware hives of the same shape and size; they are skilfully moulded with a crucifix over each entrance. Occasionally one also sees hives made of reeds, somewhat similar in shape to our own English skep, complete with hackle. But the Greek skeps are usually larger, taller and more pointed; one type, less common, has a rounded dome-shaped top. They are all more capacious than their traditional counterparts. I saw no hives of sun-baked clay or ferula stems, though *Ferula thyrsifolia* is fairly common in Greece.

In Crete, particularly on the peninsula north of Suda Bay, I saw extensive apiaries – set amidst the wild thyme – entirely of wicker hives. The bare wicker, with a few handfuls of reeds flung across the top, was all the shelter and protection provided. Some of these primitive apiaries contained more than a hundred hives.

A few miles south-east of ancient Mycenae and Agamemnon's Tomb – in Argolis, Peloponnesus – is a unique walled-in bee garth with no less than 98 bee boles, each with its basket hive, complete with the heavy coat of clay which seems traditional in that part of Greece. Even in ancient times great value was apparently placed on the direction hives should face, for each of these bee boles faces east or south-east.

The indigenous honeybee of extreme South-Eastern Europe has so far, for some inexplicable reason, never attracted any notice. True, it is not endowed with any of the glamour that would arrest attention – it lacks the bright colour and uniformity of appearance which are often so highly valued. But as a general 'business' bee, it has perhaps no equal. It resembles the Caucasian in many of its characteristics – tendency to propolise, and the construction of brace-comb. Both these defects are less highly developed in the Greek bee, and in some strains they are negligible. Its most outstanding qualities are gentleness, breeding power and disinclination to swarm. I came across no bad-tempered colonies, except in Crete. The Greek beekeeper hardly ever resorts to a smoker; a small piece of smouldering fungus is usually placed on top of the frames while an examination is in progress. The bees are as good-tempered and quiet over manipulation as the average Carniolans. Their breeding power is truly phenomenal: I am inclined to believe that no other race will equal the numerical strength of a Greek colony, or particularly of a Greek queen crossed with an Italian or Carniolan drone. But, unlike the Italian or Eastern races, breeding is severely restricted – too much so, to serve our purposes – after mid-July. The brood chamber may well be found choc-a-bloc with stores at the end of July. The brood is compact and faultless in every respect, and our experience suggests that the Greek bee is less disposed to swarming than any other race or strain we have tested in our apiaries. But it is definitely inclined to propolise and to build brace-comb freely, and the honey cappings are rather watery in appearance. Our preliminary tests and observations indicate that the Greek bee embodies the qualities required for a honey gatherer *par excellence*.

Aristotle observed that the bees of Greece are not uniform in colour; in his time the bees with yellow markings were considered best. The Greek bees of today are brown, with a yellow segment showing here and there. However, west of the Pindus range, from Messolonghi to Janina, they are uniformly black. We were assured at Janina that near Konitsa, on the Albanian frontier, a pure yellow variety could be found, but our search there revealed a mere sprinkling of yellow, which is as commonly seen east of the Pindus range as in the heart of these mountains. In these regions one rarely finds a colony absolutely uniform in colour; a small and varying proportion of the bees have one or two tawny segments. As would be expected, the queens show a wide range of colouration; drones, on the other hand, show practically none.

In Crete – according to Greek mythology the birthplace of the honeybee – the bees show a high percentage of yellow markings. Indeed, the bees of this favoured island are a 'mixed lot' in every way. Before I left Europe, I was assured that in Crete I would find also the most gentle bees extant, but the temper of some of the colonies I examined indicated a decided Eastern influence. In Cyprus I found the greatest uniformity, in Crete a deep-seated 'dis-uniformity'.

Although my experience of the Greek bee has been confined to one season, the preliminary results indicate that, given a good strain, this race may well prove to be of great value. It is definitely superior to the Caucasian, of which I had previous experience.

The indigenous bee of Western Yugoslavia, of Montenegro and Bosnia, is reputed to be more prolific and less given to swarming than the typical Carniolan of Slovenia. Though the latter has the reputation of being prolific, I have in recent years been forced to conclude that this is not so. The measure of fecundity of a race or of an individual queen is rather an arbitrary concept, and the Carniolan is undoubtedly prolific when compared with the old English native bee; Cheshire and Cowan clearly made such a comparison, and their verdict seems to have been repeated ever since without being checked. The average Carniolan is not prolific according to our standard. We have tried out more than a dozen strains recently, secured from widely different parts of its native habitat, and most of them could not fill more than seven MD frames with brood at the height of the season, whereas our own strain would readily fill ten. It was therefore with a keen concern that I looked forward to a search of the Montenegran Alps and the high mountain range along the Dalmatian coast, for I confidently hoped to find there a strain better adapted to our particular needs.

On leaving Greece I intended to make for Skoplje, then to turn westward towards Cetinje immediately north of Albania, and to go on to Ragusa, Sarajevo, Split and Ljubljana. Alas! A mishap on my last day in Greece – a burst tyre which could not be replaced – made it necessary to use the less hazardous route from Skoplje to Nish, Belgrade, Zagreb and Ljubljana.

Ljubljana, or Laibach as it used to be known, is the centre of Carniola and the headquarters of the Slovenian Beekeepers' Association – Zveza cebelarskih drustev v. Ljubljana – which helped me in my search in Slovenia. This Association, like most others on the Continent, supplies its members with all necessary equipment at cost price. It also publishes a monthly journal of very high standing, 'Slovenski Cebelar'. The members of the Association own altogether 70,000 colonies, of which 50,000 are in modern hives. The total number of colonies in Yugoslavia is about 800,000, half in modern hives.

We secured our first Carniolan queens more than 50 years ago from Michael Ambrozic of Moistrana, Upper Carniola, who founded the world-wide trade in these queens and bees. We had since then imported queens from various sources and with varying results, but it had been impossible to obtain direct imports from Carniola since 1939. I therefore looked forward with keen anticipation to visiting the central habitat of this race. Furthermore, I had an idea that I would find something of special value, apart from gaining a more precise knowledge of the environment which helped to mould the most classical type of Carniolan, which is found in this region.

Our search took us first to Lower Carniola, south and south-east of Ljubljana. The bees here are fairly uniform, but as we travelled further from Central Carniola, either east, south or south-west, the slight variations in external charactistics became more apparent. In addition, the temper of the bees occasionally left something to be desired. However, east of Ljubljana, close to the Hungarian frontier, the bees seemed to me to be more prolific and perhaps less disposed to swarming, but less uniform externally (this may be due partly to the influence of the Banet bee, a sub-variety of the Carniolan whose central habitat is further east or south-east of Maribor). A month later I had an opportunity to explore the adjoining area to the north, approaching Hungary from Styria.

The Carniolan bee in its classical form and in greatest unformity is only found in the isolation of Upper Carniola, particularly in the secluded valley running due west of Bled. The towering Karawanken to the north and north-east, the Carnic Alps to the north-west, and the Julian Alps to the west and south-west, constitute an insuperable barrier. In fact this lovely valley from Bled to Bistrica forms one of the most perfect mating stations designed by Nature, and it is not surprising that some of the best Carniolan queens are reared there. In the very centre of this valley lives Jan Strgar, known the world over as a breeder of Varniolan queens. His establishment was founded in 1903, and a considerable section of 'Slovenski Cebelar' for December 1953 was appropriately devoted to commemorating this event.

In my first report I gave a fairly comprehensive outline of the general characteristics of the Carnica bee. That description also holds good for the strains found in Carniola itself. There are

Slovenia: Jan Strgar, left, in conversation with an editor of Slovenski Cebelar. This noted breeder sent queens to every part of the world.

South-west Serbia: A novel mode of protection – the primitive wicker hives are here enveloped in slabs of bark.

PLATE X

YUGOSLAVIA

Montenegro: A modern migratory apiary near Bjelbsi, on the heights west of Cetinje the former capital.

Banat: In the acacia forests

Bosnia: A charming old-time cottage apiary.

undoubtedly some variations; indeed the wide variation between one strain and another is one of the most marked features of the race. We have had some strains which could hardly have been surpassed for uniformity in external characteristics, but which proved valueless in practice. Too much stress is often placed on uniformity, particularly in the Carniolan. There is a factor for yellow in its genetic make-up, which often manifests itself as a seasonal variation. The breeder of one of the best strains assured me that his bees will not infrequently show some yellow colouration on the first dorsal segments in the early part of the summer, but that these markings will completely vanish in subsequent generations raised at a lower temperature in the autumn. Actually the best strains (judged by performance) I have so far come across are known to manifest a fair amount of yellow. In every race, variations in colour and markings are shown in the most startling manner in the queens, and this is especially true of Carniolans. There is a danger that by placing too much emphasis on external uniformity, we may lose that much more important objective of performance.

One outstanding fact is the complete absence of brood diseases throughout the native habitat of the Carniolan bee. This impressed me deeply, for in every country I have so far visited (except Cyprus) AFB and EFB are common, and in some instances endemic. But Carinthia and Carniola seem to form an island of immunity. Acarine, Nosema, and paralysis are present, but not foul brood. Its absence cannot be fortuitous (the mountain barriers would retard, but not prevent, the spread of disease, and I have seen AFB in an almost inaccessible region of the Pindus mountains on the fringe of Albania). We are dealing here not with a true immunity, but probably with an innate resistance.

Beekeeping conditions in Carniola, especially in Upper Carniola, are very similar to those in the adjoining Austrian Province of Carinthia. However, in Central and Lower Carniola, especially in the mountainous region along the Adriatic, there is a more varied nectar-bearing flora. In Upper Carniola honey-dew from the pines forms the main source. In Central and Lower Carniola limes abound, and they seem to yield freely here; they were in full bloom at the time of my visit, and I was able to sample pure lime honey. Another honey of high quality is gathered in August and September in the mountainous region of Dalmatia, from the mountain savory, *Sattureia montana*. Some of the more enterprising professional beekeepers transport their colonies in spring to the rosemary, which grows in great profusion on some of the islands off the Dalmatian coast. Some wonderful crops of a honey of supreme quality are thus secured. Many colonies are also moved into the Istrian Peninsula at the end of June, for the honey from the sweet chestnut which is, however, of a lower quality. There are many secondary sources, and the flora in general is more favourable to beekeeping in north-west Yugoslavia than in the adjoining Austrian territory.

I have no idea when house apiaries first came into use. In Carniola beehouses are an accepted and integral part of both primitive and modern beekeeping. For migratory beekeeping the hives are stacked in sectionally constructed sheds. I did not see any beehouses in Yugoslavia outside Carniola.

LIGURIAN ALPS

On leaving Yugoslavia I had a number of enquiries to make in the adjoining Carinthia and Styria, which in due course proved of great value. However, the Ligurian Alps were the next important sphere of search. A brief visit had been made there in October 1950, but we were unable to secure any Ligurian queens as the season was too far advanced.

The world-wide fame of the Italian bee is partly based on the success achieved with the first imports made nearly a hundred years ago. These bees came from the Ligurian Alps – hence the name Ligurian bee. Our findings indicate that the genuine leather-coloured Italian, which embodies all the desirable qualities which have made the Italian so popular, is only found in the Ligurian Alps, in the mountainous region between La Spezia and Genoa.

Apart from the direct practical value, I felt that a more precise knowledge of the tawny Lugurian would have a great bearing on our future cross-breeding experiments, and after much effort I now succeeded in securing queens of the type required. The parcel containing the collection of previous queens was left overnight in my room, ready for posting next day. To my amazement, the next morning both table and package were covered with tiny black ants, and on touching the parcel, thousands of these wretched creatures fell out of the cotton wool packing surrounding the cages. All the queens and bees had been killed by the ants. The loss of the Ligurian queens proved the greatest disappointment of my journey: I could not retrace my steps, for the required time and energy were no longer at my disposal.

However, I went on to the South of France in the firm belief that I could include the Iberian Peninsula. But it soon became clear that the long sustained effort since February called for a halt and an overdue rest, and I returned to Buckfast on September 29th.

SUMMARY

Gradually, but surely, and step by step, information of value concerning the manifold races of the honeybee is being accumulated, and a more precise knowledge of the range of their distribution is emerging. The jig-saw puzzle of the races can thus slowly be pieced together. The mode of their evolution is being revealed stage by stage, so that the individual defects and qualities can be more readily traced to their primary sources. We are by degrees coming to a truer and more perfect understanding of the vast fund of potentialities which is at our command for the creation of the 'perfect bee'. But much remains to be done, for in an undertaking of this nature, where unforeseen difficulties and delays are inevitable, time is an all-important factor.*

*In 1954 and 1956 Br. Adam undertook two short journeys, one embracing the northern half of Asia Minor and a number of the Aegean islands. The other one took him to Yugoslavia: Bosnia, Servia, Montenegro and Hrzogovina. The results and findings made on these journeys are included in the final report of 1962.

1959
THE IBERIAN PENINSULA: SPAIN – PORTUGAL

At the beginning of September 1959 Br. Adam set out for the Iberian Peninsula. He entered the Peninsula by way of the Mediterranean end of the Pyrenees and left two months later on the Atlantic side via Irun. In the interval he covered no less than 6,500 miles by car, travelling from Gerona in the extreme north-east to Lagos in the extreme south-west, and from Tarifa – the most southerly point – to Coruna in the north-west corner of the Peninsula. It proved an arduous journey, but one well worth the effort. He was able to collect queens from every part of the Peninsula and samples of bees in still greater number for the biometric studies.

SPAIN – PORTUGAL

Dr. F. Ruttner (1952) points out that during the ice-age – which extended over a million years – climatic conditions were such as to preclude the existence of the honeybee from the greater part of Europe. The great Scandinavian ice-sheet extended from the North Cape as far south as a line from the Severn Estuary in England in the west to Kiev in Russia and beyond in the east. The Pyrenees and Alps were covered in glaciers, and the country extending northwards to the fringe of the Scandinavian ice-sheet presented a vast tundra. The fossil remains so far discovered in Europe all date from the Tertiary period. During the ice-age the European honeybees were left with three places of refuge on the Continent: the Iberian, Appennine and Balkan

Peninsulas. The bee of the Appennine Peninsula, the Italian bee, was probably always confined to its country of origin, for at all times the Alps formed an insuperable barrier to any migration northwards. On the other hand, after the ice-age the bee of the Balkan Peninsula could spread northwards as far as the eastern barriers of the Alps, and north-east to the fringes of Southern Russia where its further progress was seemingly checked, not by mountain ranges, but by vast treeless steppes. The re-population of the rest of Europe, subsequent to the last glacial period, was therefore left to the bee of the Iberian Peninsula. The gap at both ends of the Pyrenees permitted a northward migration without let or hindrance. This post-glacial return of the honeybee to Central Europe took place about 7,000 years ago.

In view of the fact that the black European bee emanates from the Iberian Peninsula, Dr. Ruttner holds it should be named after its land of origin, just as the other two European varieties have been named after the countries where they are now found in their most typical form. Whilst there can be no doubt that the black or brown European bee – and, indeed, all the bees found throughout Northern Russia – originate from Iberian stock, it is certain that the Iberian bee in turn – in the still more remote past – originated from the North African bee, commonly called Tellian, or *Apis mellifera unicolor* var. *intermissa*. In my report published in 1954 I expressed the view that the Tellian was a primary race, and that the numerous varieties of brown or black bees – at least those of Western Europe – evolved in the course of time from the Tellian.

I also pointed out that I had not as yet had an opportunity of exploring the Iberian Peninsula, but that in the strains found in Southern France and North-Western Europe the variation from the prototype was only a question of degree. The close relation was obvious. The pattern of evolution north and north-eastwards from the Pyrenees could be readily traced, and the differences were only a matter of intensity and degree. It was clear to me beforehand that although the Iberian Peninsula represented a 'halfway house' in the stages of development, it was nevertheless the vital link between what is termed the European black bee and the prototype. For all we know the glacial and inter-glacial periods extended over a million years, and as recently as 5,000 BC. *Apis mellifera* var. *mellifera* was confined to the territory south of the Pyrenees.

Here it was also virtually isolated from any contact with the African continent and still more completely from the rest of the world. The Straits of Gibraltar at their narrowest point are 9 miles in width, and it may be safely assumed that no swarm would cross this distance on the wing. The almost constant high wind from the east, confined to the Straits and immediate neighbourhood, would render the crossing of a swarm doubly impossible.

Apart from these considerations I looked forward to a better acquaintance of the bees and beekeeping in the Peninsula, for I was already in possession of a great deal of information on these subjects. The information was obtained from a young Spanish monk who stayed at Buckfast from 1926 to 1928 to learn beekeeping. He belonged to the Abbey of Valvanera in the north of Spain; bees and beekeeping have been linked with this Abbey in the hearts of Spanish beekeepers in a very special manner since Our Lady of Valanera is regarded as the Protector of beekeepers throughout Spain. This young monk, along with 18 other members of his community, was alas, killed in the Civil War in the autumn of 1936.

The Iberian Peninsula is a world to itself in many ways. It is cut off from the rest of Europe by a mighty mountain barrier, difficult to cross except at its extremities. It is also a land abounding in sharp contrasts. To the south-east and north-west there are mountain ranges with Alpine grandeur rising above the snow line. Amongst these mountains can be found tucked away rich and lovely valleys. By contrast great central plain or *meseta*, with an average altitude of 2,000 ft., presents a huge expanse of dreary uniformity and extremes of temperature – a furnace in summer and a refrigerator in winter. The eastern fringe along the Mediterranean is blessed with an equable climate, with no winter in the strict sense of the word. On the western seaboard, northward from Lagos to Coruña, the moisture-laden winds from the Atlantic penetrate many miles inland, and cause this section to be exceedingly fertile. Southern Spain and Portugal, particularly Andalusia, have warm winters and torrid summers. The distribution and type of rainfall presents contrasts as striking as does the country itself. The northwest of the Peninsula has an average precipitation of 24 inches and over, with 66 inches at Santiago de Compostella – equalling our annual rainfall at Buckfast – and 12 inches or less in the southeast of Spain, but 35½ inches in the Gibraltar area. The rains in the north-west are of the same type and intensity as those experienced here in England. The day we spent at Vigo, and some weeks later in the north of Portugal, the rain was every bit as persistent and torrential as we are accustomed to in Devon. In the arid sections of Spain the rains are confined to the autumn and winter, but are spasmodic and very uncertain. They then come in short, sharp downpours often seemingly out of a blue sky. Downpours of this kind cannot penetrate the hard-baked crust, but merely tend to wash away any fertile surface soil. When the rains do not materialise, which happens all too often, poverty and want are the result.

Owing to the extraordinary diversity of climate, altitude, exposure and soil, the Iberian Peninsula is richer in plant species than any other section of Europe. Endemic species are particularly numerous. The most typical trees of the arid parts are the two species of oak, the evergreen holm-oak *(Quercus ilex)* and the cork-oak *(Quercus suber)* and, of course, the carob *(Ceratonia siliqua)*. On the *meseta* the main roads are often lined with *Robinia pseudoacacia*, which is about the only tree to be seen here. The predominant vegetation of the *meseta* and the stony uncultivated areas, of which there are immense tracts everywhere, is a scrub of evergreen bushes and herbaceous plants of the families Cistaceae and Labiatae. To the latter belong thyme, lavender, sage and rosemary – the great nectar-bearing plants of the Iberian Peninsula. Gorse and broom, the Spanish broom *(Spartium junceum)*, are extremely abundant in Galacia in the moist north-west, together with many species of *Erica*. Indeed, there are extensive tracts of moorland in the mountainous area on a line north-west from Braganza to Bilbao. *Calluna vulgaris* seems, however, very much more common in the mountainous parts of Northern Spain and in the woodlands of this region. I came across the first ling in bloom in a copse between Almazan and Soria, then next day far greater areas on the way to Logroño. It is also

Experimental apiary of the Agricultural Station of Murcia.

Orange blossom. The orange groves of Valencia provide one of the richest nectar sources found in Spain.

PLATE XI

SPAIN

A type of beehouse, beside the road between Burgos and Vitoria, reputed to date from Moorish times.

Wickerwork hives provided with a coat of clay and whitewash in the neighbourhood of Soria, Old Castile.

A cork hive of Andalusia such as has been used from time immemorial.

93

The jungle type of vegetation in the clearings of the cork forests of southern Portugal furnishes untold sources of nectar, amongst them Calluna vulgaris which attains here a height of more than 4 ft.

PLATE XII

PORTUGAL

Portugal and Spain possess extensive cork forests and huge stacks of cork are a common sight in summer.

A row of cork hives with entrances at varying heights.

commonly found, particularly in Southern Portugal, amongst the undergrowth of cork woods. Here the ling flowers considerably later than in Northern Europe, and it is not stunted and gnarled as with us; its growth is tall and the flower spikes are formed in elongated sprays. A seemingly endless number of *Erica* can be found widely over the Peninsula. The most common are Spanish heath *(Erica australis)*, Portuguese heath *(E. lusitania)*, *E. arborea alpina*, a native of the Spanish mountains, *E. umbellata*, and *E. scoparia*.

Eucalyptus is fairly common in Andalusia and parts of Portugal. In the Province of Huelva I observed extensive plantations of a good many year's standing. Two of the most common are *Eucalyptus globulus*, flowering in November-December, and *E. rostrata* which flowers from mid-June to mid-July. The latter yields nectar only in the evening and early morning hours. The great orange groves are confined to a relatively restricted area, south and north of Valencia and west of Sevilla. The Spanish chestnut *(Castanea sativa)* I observed in greatest abundance in the north of Portugal, in the area between Braga, Vila Real and Braganza. White clover *(Trifolium repens)*, though common in North-West Spain, is not valued as a source of nectar. In Andalusia large areas of cotton *(Gossypium herbacceum)* are grown, but poison sprays often cause heavy losses of bees.

Clearly the Iberian Peninsula is blessed with a superabundance of nectar-bearing trees, shrubs and plants, the most important of which are undoubtedly orange, rosemary, lavender, thyme, ling and the various *Erica*, *Eucalyptus* and perhaps the carob.

All these details may seem perhaps beside the the main point and purpose of my search. However, I would stress that one of the primary objects of a journey such as this is to acquire an intimate knowledge of the history and origin of a race of bees, as well as the environments and influences which have been at work in the formation and development of a particular race and strain. It must be remembered that the habitat in which an organism has been formed and moulded in the course of time bears a close relation to the characteristics with which it is endowed. Indeed, the particular characteristics of an organism often mirror the particular influences of its habitat, and there is perhaps no organism for which this is more so than the honeybee. In nature the bee is at the absolute mercy of its environment, and must either adjust itself to it or perish.

On my entry into Spain I endeavoured to explore as best as I could the north-eastern corner of the Peninsula before proceeding to Madrid. The Province of Catalonia, with its varied flora and relatively humid climate, is a good region for beekeeping. The Layens hive of French origin, is in fairly common use here. There are no supers to this hive; the capacious brood chamber, holding 14 frames measuring $13\frac{3}{4} \times 11\frac{3}{4}$ in., provides room for both brood and surplus stores. It is a chest-like construction, with a flat roof hinged to the body, and both ends of the body fitted with chest-handles. Its great advantage is ease of transportation – an important consideration where migratory beekeeping is an acknowledged routine. This holds good for a great part of the country adjoining the Mediterranean. After rosemary and orange have finished flowering, the hives are transported to the higher regions of the central plateau, where thyme abounds in June and July, and also lavender and sainfoin here and there. Rosemary offers a small second crop along the coast at the end of September. As I had travelled south of Narbonne a few days before, through the world-famous rosemary district of Corbières, I noticed that it was just coming into bloom again. In Catalonia the surplus honey yield averages about 25 kg. per colony.

According to the best informed sources, there are about 1,200,000 colonies of bees in Spain, of which one-third are in primitive hives. But the actual number may well greatly exceed this figure. Portugal, which covers only 15 per cent of the Iberian Peninsula, has a total of 473,642 colonies, of which 111,924 are in modern hives. The relative density of colonies per square mile is therefore approximately 13.9 for Portugal and 6.5 for Spain. The significance of these figures can perhaps be better appreciated if they are set against an average of 3.8 for England and Wales, with a present-day total of 219,545 colonies.

In both countries the Langstroth hive is the one in most general use. Indeed, the catalogue of the foremost appliance firm in Spain offers only the Colmena 'Perfection' (Langstroth) and the

Layens hive. No shallow supers are used, only full-depth Langstroth bodies as supers. Two firms specialise in the manufacture of comb foundation, one at Alcira (Valencia) and the other at Andujar (Jaen).

Primitive beekeeping is with good reason still firmly entrenched in both Spain and Portugal. In Leon and Orense I came across log hives, and in Old Castile some made of wickerwork, complete with the usual coating of clay. However, cork is the material usually employed for the construction of primitive hives in this part of the world. The extensive cork forests furnish a material which is ideal for the purpose, especially as it is such an excellent insulator. The cork also costs next to nothing, and entails little skill or effort in making up. A sheet of cork, as ripped off the tree, is allowed to resume its natural shape, and a few pins of cistus wood are driven across the vertical joints to hold it together; a flat section of cork is fitted over the top end of the cylinder, forming the roof, and the hive is ready for use. No special skill is required, as in making wicker hives, or our own straw skep. Columella tells us that in Roman times the making of cork hives was an occupation of slaves in their leisure time.

The diameter of cork hives varies somewhat, but is usually about 10 in.; the height is about 27 in. The hives are invariably used in an upright position (never horizontally and in tiers, as is customary in Sicily and the Middle East), and usually in great numbers. It is no uncommon sight to see a 100 or more of these cork hives in one place, lined up in rows one behind the other. Indeed, these old-time beekeepers have a saying: 'De cien uno y de una cien', meaning out of a hundred, one, and out of one, a hundred – an allusion to the transitory character of colonies in unfavourable years and their magic increase in good seasons and favourable circumstances.

It may perhaps surprise many to hear that in Spain and Portugal beekeeping is carried out on as large a scale as anywhere in Europe. Indeed, with an average colony density of about 7.5 per square mile, beekeeping must necessarily play an important role in the life of the country. However, there is no intensive beekeeping as we know it. The beekeeping here is of the let-alone type, with commercial apiarists relying on migratory beekeeping for remunerative returns. There is virtually no queen rearing and no effort is made to improve the stock. Italian queens are imported here and there, but not on any extensive scale. But large commercial beekeepers are often found in most unexpected places. I came across one between Zamora and Salamanca with 800 colonies. Near Sevilla there is an old-established family concern with 2,000 colonies, operating a honey-packing plant equal to any in Northern Europe. This firm puts up its honey in most attractive ornamental jars of different sizes and designs.

Beekeeping throughout Spain comes within the scope of the Veterinary Service. At the Provincial Agricultural Stations beekeeping is usually represented; modern apiculture is also taught at the larger Agricultural Colleges, of which I visited a number. One in the south, near Cape Trafalgar, operates no less than 6,700 acres and teaches every branch of agriculture, including beekeeping. Another college, near Zamora in the north-west, seemed to me equally comprehensive. These are privately-owned colleges, not State owned. I came away with the impression that the central authorities in Spain are not greatly interested in furthering beekeeping. There is, however, a movement on foot to get a national bee research institute established, but whether anything will come of it remains to be seen. It certainly seems a pity that beekeeping is not receiving the help it needs, for good advancements are undoubtedly possible in every direction.

Conditions in Portugal are somewhat different in this respect. From the precise statistics as to the number of colonies there, it may be assumed that beekeeping occupies a more favoured position. Sr. Vasco Correia Paixao is technical beekeeping adviser to the Ministry of Agriculture. He is also in charge of the Posto Central de Fomento Apicola. I observed much practical evidence of the solicitude of the Ministry to help beekeeping. The University of Oporto has published an extensive treatise on the pollen analysis of Portuguese honeys (Martins d'Alte, 1951).

It seems rather surprising that no comprehensive survey or study of the honeybees of the Iberian Peninsula has hitherto been attempted. I have already pointed out that there can be little doubt that it is from this stock that all the dark races of *Apis mellifera* have sprung, and that the Iberian honeybee is in its turn descended from the Tellian. The possibility that the Iberian stock was the original one, and that the migration from the Peninsula was to both north and south, cannot be maintained, because it is the Tellian bee which possesses in the highest and most concentrated form all the characteristics manifested in the numerous sub-varieties.

Since the honeybee pays no heed to political or national frontiers, it seems hardly correct to speak of a Spanish or Portuguese bee. Nor can there be any question of several races, for there are no mountain barriers which could isolate one section of the Peninsula from another, and so allow the development of a distinctive race. There are, however, a number of distinctive strains, and this seems most likely to be due to local environments caused by the widely differing geographical and climatic conditions in the Peninsula. This supposition has good evidence to support it, but it must be emphasised that the differences are never more than those of degree of intensity of the basic characteristics. Just as it is wrong to expect something not present in the prototype, it is equally wrong to suppose that widely differing geographical and climatic conditions would have no selective effect on the basic characteristics, especially with such a susceptible creature as the honeybee.

Luis Méndez de Torres in his treatise on beekeeping, published by him at Alcalá de Henares in 1586, speaks of the great diversity in size, temperament and colour of the bees of his day. This certainly holds good today. But the great diversity is not confined to size and temperament; it extends to all the qualities on which performance is based. The Iberian honeybee is in the main jet black, and the blackness is often accentuated by scanty tomenta and overhair. Nowhere could I observe bees that could be described as yellow, except recent importations. I did, however, now and again observe clear yellow markings, confined to the area where the first three dorsal segments join the ventral plates, as similarly noted in the Tellian bee here and there in North Africa. The queens are black and very uniform in colour; they are quick in movement and rather nervous. They are prolific, but their fecundity is largely controlled by the 'wherewithal' or the lack of it. In other words, they do not breed to excess in times of dearth, as Tellians are disposed to do. Such a disposition would often mean death from starvation during the long spells of dearth. On the other hand, a commensurate fecundity is at hand, and is given full scope when conditions warrant. A flexibility of this kind is essential in the climatic conditions found on the Peninsula. Colonies can build up to huge populations when conditions are right, and the economic value of such colonies is here safeguarded by a moderation in swarming. The extreme addiction to swarming of the Tellian is its undoing from the practical beekeeper's point of view. The Iberian bee shares with the Tellian in undiminished form its extraordinary hardiness. It is active – and to good purpose – at temperatures when other bees would not venture forth. It shares also in full the susceptibility to brood diseases, the lavish use of propolis, and the watery appearance of the cappings. The extravagant use of propolis is one of the most undesirable traits of the Iberian bee. However, strains can be found which for all practical purposes do not show this fault. As to temper, the bees of North-Eastern Spain and those along the foothills of the Pyrenees seem more irritable than those in the rest of the Peninsula. But I came across some really bad-tempered colonies in widely varying regions, for instance south of Malaga, and again north of Lisbon. On the whole the Iberian bees are certainly not as good-tempered as the Italians, but by no means as aggressive as many French bees.

These observations are based on what I saw when in Spain and Portugal, and on the experience at Buckfast confined to the 1950 season. As that summer proved a complete failure from the end of June onwards, no comparative results could be obtained as to the honey-gathering ability of the pure Iberian bees or first-cross hybrids. A number of seasons must elapse before trustworthy results are available. I would, however, be greatly surprised if the Iberian did not match up to the French bee, which experience has shown to be the most outstanding honey gatherer of all the European races.

I referred already to the Iberian bee's susceptibility to brood diseases, a defect which nearly every variety of the common European dark bee is subject to. Susceptibility to acarine disease is another characteristic defect shared by all the varieties which have evolved from the Tellian, their common ancestor. Acarine disease is very prevalent in the Iberian Peninsula, particularly along the Mediterranean seaboard and in Andalusia. Indeed, I was informed that the losses were so great as to have caused a serious decline in the colony population of Spain. The authorities have come to the conclusion that remedial measures are of little value, and that the development of a bee resistant to acarine offers the only long-term solution. Experimental work on these lines is being conducted at Malaga.

CONCLUSION

When I set out for Spain I confidently hoped I would have an opportunity to see the unique rock-shelter painting near Bicorp, a little more than 50 miles south-west of Valencia. This well-known rock painting in the Cueva de la Araña depicts a person, up on a rock face, taking honey from a cavity in the wall. It is the oldest existing record of its kind connected with beekeeping, its age being variously estimated from 8,000 to 10,000 years. It was in all probability painted at a time when most of Europe north of the Pyrenees and Alps was still held in the last lingering grip of the ice-age.

We left Valencia early in the day, but other tasks prevented us from reaching Bicorp before 4 p.m. – only to be told by the local folk that the rock shelter was a good hour's walk away and that it could only be reached on foot. We had not got the necessary time at our disposal, for we had to reach Alicante that same day, which was no small distance from Bicorp. To our grievous disappointment, we were left no other choice but to depart without seeing the rock painting.

During the early part of my journey I had to put up with great extremes of heat. I shall always remember the day spent at Murcia, when even my companions – who were used to high temperatures – found the heat almost beyond endurance. Towards the end of the journey heavy rain greatly impeded our progress in the north of Portugal. It turned cold, too; we had hardly finished the inspection of the last apiary, situated on a ledge of a near vertical mountainside overlooking Colvilha, when a violent hailstorm made us run for shelter. Next morning, when I set out on my return journey from Guardia, it felt definitely wintery. Thanks to the determination of my assistants the task was completed in the nick of time.

1962
MOROCCO – TURKEY – THE HONEYBEES OF ASIA MINOR YUGOSLAVIA: BANAT – THE AEGEAN ISLANDS EGYPT – THE EGYPTIAN HONEYBEE IN THE LYBIAN DESERT

The concluding reports bring not only the impressions and experiences which Br. Adam collected on his last expedition in 1962, but also take into consideration the results of two shorter journeys made in 1954 and 1956. In the autumn of 1954, Br. Adam covered the northern half of Turkey and, subsequent to his visit to Crete in 1952, included on this occasion a number of the more important islands in the Aegean. In a paper entitled 'The Honeybees of Asia Minor' he submitted to the International Beekeepers' Congress of 1958 in Rome a preliminary report. In July 1956 he was able to visit Western Yugoslavia, namely, Bosnia, Herzogovina and Montenegro – a part of the country he intended to include in his trip of 1952, but was prevented from doing so on account of mechanical trouble with his car.

IN SEARCH OF THE BEST STRAINS OF BEES

The final journey

Br. Adam terminated his great undertaking in 1962, again by an extensive journey made in two parts. On March 26th he left England with Morocco as his first objective. A well organised expedition took him through that country to the fringe of the Sahara and to the Tafilalet, a group of oases of great importance to his search. On the way north from Marrakesh he traversed virtually the whole length of Morocco. At the end of April he left by sea for Asia Minor and on this occasion covered the whole southern half of Turkey. The northern half he had visited in 1954. On completing his work in Turkey, he set out for North-Eastern Yugoslavia in search of the Banat bee. At the end of June he returned to England to attend to various urgent tasks. On October 23rd he left by air for Cairo. In Egypt his search was confined to the regions of the Nile Valley and Delta and to a few of the more important cases situated in the southern sections of the Lybian Desert. In these oases the Egyptian Ministry of Agriculture operates a number of breeding stations. After finishing this last task, he returned to Buckfast early in January 1963.

MOROCCO

I crossed from Harwich to the Hook of Holland on the night of March 26th/27th to avail myself of the Autobahn from the Hague to South Germany, and from there followed a route via Lyon, Narbonne, Barcelona and along the Mediterranean coast to Gibraltar, where I awaited the arrival of Dr. R. H. Barnes, who volunteered to accompany me on the journey to Morocco. I heard his plane coming in shortly after midnight, and we met at breakfast the following morning. A few hours later we were on our way to Tangier.

I had firmly intended to visit Morocco in 1952, but when I was in Algeria I was prevented by various difficulties from proceeding westward to the adjoining country. On looking back, I feel this delay proved fortunate, for I would most certainly have never been able to carry out the work to my satisfaction in the circumstances then prevailing. I was not specially interested in the native black bee of Morocco for I realised it could not differ materially from the indigenous bee of Algeria, A. *mellifera* var. *intermissa*. The primary object of my visit to Morocco was to obtain more precise knowledge of the Saharan bee and its habitat. In this connection Monsieur P. Haccour of Sidi-Yahia du Gharb, whom I met at the Congresses in Rome and Madrid, rendered me invaluable service. Mr. Haccour, who owns about 2,000 colonies, is one of the keenest commercial beekeepers I have had the pleasure of meeting. He spoke Arabic and possessed a life's experience in dealing with the Moroccan people.

So our first point of call was his home, a country house some miles from Sidi-Yahia, set in the midst of eucalyptus, mimosa, citrus and many other kinds of sub-tropical trees. The heavy scent of orange blossom pervaded the area, particularly early in the morning before the sun dispelled the high humidity; by noon the temperature approached 90°F (32°C). We arrived at a season when the countryside was attired in its richest flora. An exceptionally heavy rainfall during the previous few months had made the flora unusually luxuriant. After two days in this marvellous setting, spent in visiting some of the beekeepers nearby, we set out for the Sahara with Mr. and Mrs. Haccour.

Our route took us across the Northern Atlas, via the Col du Zad. Here at 6,600 ft. we were back in wintery conditions, surrounded by snow everywhere. Indeed, we were told that we would not have got through this Pass by car a week earlier. The night was spent at Midelt, a small village in the eastern foothills of the Atlas. Next morning, as we approached the fringe of the Sahara, the character of the vegetation changed, and date palms made their appearance here and there. In place of the bare rock and boulders, sand dunes came in sight. Well before midday we reached the Rafilalet, a group of oases which Mr. Haccour considers to be the cradle of *A. mellifera* var. *sahariensis*.

I believe it was Ph. J. Baldernsberger who first drew attention to the race in 1921. He discovered this bee at Fuguig, the most easterly Moroccan oasis. As far as our present knowledge indicates, Figuig is also the most easterly point at which the race can be found. It is certainly not found in the more well-known oases of Algeria, such as Laghouat, Bou Saada, Biskra or Ghardaia. Westwards its distribution extends at least as far as Quarzazate, as we were able to verify ourselves. It should be realised that this race is hemmed in by two great natural barriers: by the towering Atlas chain of mountains on the north-west and by an endless waste of sand in the east and south. Moreover, each of the various oases is

Part of a primitive apiary of about 300 Moroccan type wicker hives, scattered about on the ground with wanton abandon.

PLATE XIII

MOROCCO

Tafilalet

In the Rif: Mr. Haccour selecting the colonies near Torres de Alcala.

A cork hive wrapped in a jute sack, ready for transportation to Targuist.

almost as effectively isolated one from another by miles of barren desert extending between them. As far as I could ascertain, there can be little or no inter-breeding in most localities.

The question arises: how did this race originate? There can be no doubt that this Saharan bee is a distinct race – distinct in its external and physiological characteristics. We know that throughout North Africa, from Tripolitania to the most southerly point of Morocco bordering the Atlantic, the jet-black bee, *A. mellifera* var. *intermissa*, holds undisputed sway. But here, wedged in between the Atlas and the Sahara, we have within a relatively small area, confined to the fringe of the desert, miniature pockets of a distinct yellow race of bee. I cannot for one moment believe that *sahariensis* should in course of time have evolved from *intermissa*. There is no similarity between the two races. Mr. Haccour holds the view that Jewish immigrants may have brought the original stock from the Middle East more than 2,000 years ago, and that in the intervening years, due to the special environment, the bee we now know as *sahariensis* was evolved. However, all the Middle East races are well known to me, and I can discern little or no similarity. Externally, *sahariensis* resembles *Apis indica* more than any other bees, but the similarity extends no further.

The pure *sahariensis* is not yellow; the colour might best be described as light tan. But a wide variation is manifested, and the colour extends in varying degrees to all the dorsal segments. Owing to the darker colour and the considerable variation in markings, the Saharan bee is by no means as attractive as the more brightly coloured races. In size this bee is midway between *ligustica* and *syriaca*. The queens also vary in colour, from bright yellow to dark brown – though never black. The drones are remarkably uniform and have two conspicuous bronze-coloured segments.

I have found the pure queens moderately prolific. The bees are relatively good tempered, but rather nervous, particularly in times of dearth. When a hive is opened they run to and fro, just as wasps will when their nest is disturbed. They also fly up in great numbers, but do not act aggressively. Also, when under manipulation, the bees fall off the combs very readily. They seem to have the least foothold of any bees I know. In this respect the Italian bee represents the other extreme – she can only be dislodged with force. One other notable characteristic of *sahariensis* is its quick flight from the entrance. There is no loitering of any kind – a quality which I believe Baldensberger already noted. She tends to propolise, but not excessively. The pure *sahariensis* suffered a heavy loss of bees at Buckfast in the severe winter of 1962/63, but the colonies survived in surprisingly good condition and strength. Those with first-cross queens wintered outstandingly well in every way.

A first-cross, Saharan queens to our own drones, has proved surpassingly prolific – indeed, the most prolific cross we have so far tested in our apiaries. In addition, the brood is wonderfully compact and – most remarkable for a first-cross – little or no drone brood is raised. This characteristic was manifested by every colony with a first-cross queen of this type. I regard it as a most desirable quality, for most hybrids tend to raise drones to excess, and certain crosses will invariably spoil a set of combs or foundations to such an extent that their further use is uneconomic. Though the pure *sahariensis* is reputed to be addicted to swarming, I have not found this to be so in a first-cross. It is premature to express a view on the nectar gathering and general foraging ability of this cross, since the summer of 1962 proved a complete failure in south-west Devon. Indeed, it was the worst season in my 49 years of beekeeping. I will, however, say this: the Saharan bee, when suitably crossed, has great possibilities. On the other hand, the pure *sahariensis* is itself unlikely to prove of much value to the beekeeper.

A number of claims are made for this race, such as exceptional tongue reach, great wing power and foraging ability. The question of tongue reach will be determined as soon as reliable biometric data are at hand. The *sahariensis* is undoubtedly an exceptionally active bee, but I cannot say whether her range of flight is as great as has been assumed. Evidence may be forthcoming later to give us some reliable information on this point. Considering the environment in its native habitat, the assumption may well prove correct.

One of the very first things that struck me on arrival at Erfoud, the principal town in the Tafilalet, was the seared and ragged condition of the palm trees. They looked withered and lifeless, with none of the deep green we usually associate with the fronds of a palm, and which I had seen in the Algerian oases and in other parts of the world. These palms gave an indication of the climate and environment in which the Saharan bee ekes out an existence. Here, on the fringe of the Grand Atlas and the Sahara, temperatures vary from near freezing in winter to nearly 120°F (49°C) during torrid spells in summer. There are sharp differences between night and day temperatures in all desert countries, but here they seemed especially harsh.

Apart from a few desert flowers, the date palm, eucalyptus, citrus, lucerne and various other legumes furnish the main sources of nectar. The legumes are cultivated in small plots in amongst the palm trees. From the condition of the colonies I was able to inspect, I could only conclude that the fight for survival here is of the most exacting kind. Where a live colony should have been, all too often we found an empty hive and traces of the erstwhile combs. The number of colonies in the various oases we visited seemed small at best. It is therefore not surprising that the local beekeepers will not readily part with a queen, or still less with a whole colony.

In my 1950 Michelin map most of the regions we passed through were marked as zones d'insécurité, and modern beekeeping has not had time to penetrate in to these out-of-the-way places. (We came across only one modern hive, in the gardens of the Governor of Goulmina.) The bees are kept, according to custom, in cavities in house or garden walls. The walls are constructed of sun-baked clay, and the cavities are not very spacious – usually about 8 in. high, 10 in. deep and about 20 in. wide (20 × 25 × 50 cm.). Access to the cavity is obtained by removing a wood cover (made either in one piece or of a number of individual boards) which is cemented in place with clay. Where the cavity is in the wall of a dwelling, access is obtained from inside the house or a room. This seems the most common way of beekeeping in these remote places. However, at Goulmina in the gardens of the Governor, I observed a number of special constructions in clay of unusual dimensions and design. The entrances were fitted with a guard to prevent the intrusion of marauders – a board about 8 in.

square fitted with auger holes of a size permitting the passage of a bee, but nothing larger. This is apparently a necessary precaution, though at the time I could not perceive any evidence of the many enemies that exist in other parts of North Africa, apart from the wax moth.

I had observed no such dearth of bees in the Algerian Sahara. At Laghouat for instance – an oasis no larger than those visited in Morocco – there were at least 50 colonies: of black Tellian bees, of course. Admittedly, in the Moroccan oases there is no beekeeping; bees are merely housed and left alone. Now I have a measure of experience with the Saharan bee in England, I can only ascribe the scarcity of bees in its native habitat to a combination of exeptionally adverse circumstances. Indeed, it seems difficult to believe that this race should have been able to establish itself at all in such a environment, and survive to our day.

We were unable to include in our search the oases east of the Tafilalet, but went west from Ksar-es-Souk as far as Quarzazate. We can say that *sahariensis* extends from Figuig to Quarzazate, but the actual limits of its distribution east and west of these points remain undetermined.

From Quarzazate we crossed the Southern Atlas by the Col du Tichka (7,448 ft.), passing to our left the Dj Toubkal (13,644 ft.), the highest mountain in North Africa. All the way from the Tafilalet to Quarzazate we were rarely out of sight of snow-capped peaks. Now we were in the midst of snow again, but not for long; in another 74 miles Marrakesh was reached, from where we retraced our way northwards again.

The primary objective of my visit to Morocco was to gain a first-hand knowledge of the Saharan bee and its habitat, but I also took the opportunity to extend my knowledge of the black North African bee found in the regions west of the Atlas. It was soon apparent that the French colonists had at one time or another imported queens from Italy, and possibly from America. Even south of Marrakesh indications of these imports could be observed. In general the black indigenous bee did not materially differ from the Tellian bees as found in Algeria – with the one difference that their temper, which was bad enough in Algeria, had here developed to a savage ferocity. I came across one exception near Petitjean, at a rather remote apiary of about 300 colonies belonging to a Berber family. Their bees resembled more closely in external appearance the Carniolan bee, and could be handled with a measure of impunity. If these colonies had been in modern hives, owned by Europeans, or within a few miles of a village or town, I would have concluded this was the result of an importation, but the apiary was far from any habitation; the owners lived in Bedouin-type tents; the hives were of wickerwork and lay abandoned amongst weeds and grass. And to complete the primitive picture, there was a skull of some creature hung up to ward off evil.

On our way northward from Marrakesh we traversed almost the whole length of Morocco. Due to the exceptional rainfall during the previous winter, the country was a riot of colour. Shortly after leaving Marrakesh, and the last of the palm groves, we came to a veritable ocean of yellow extending as far as the eye could see, seemingly of the common mustard (*Brassica campestris*). A little further on were extensive areas of coriander (*Coriandrum sativum*), cultivated for its fruit. Bees were working the coriander vigorously. Presently great expanses of the North African marigold came into view. Most of the northern half of Morocco west of the Atlas was a like a vast bed of flowers, with a greenhouse temperature and humidity. From all I could tell, this region must offer great possibilities to an enterprising apiarist.

ASIA MINOR

As the Karadeniz entered the Dardanelles by the first light of day on April 23rd, my thoughts went back to the time of the First World War. The ridges to the left, for which men fought so fiercely, were covered in spring-time flowers and glowing in the warmth of the rising sun. The mainland to the right was, I knew, regarded as one of the most favoured regions for beekeeping in Asia Minor.

I first visited Turkey in the autumn of 1954. On that occasion I came by road through Yugoslavia and Northern Greece. Eight years ago the road from Istanbul to Ankara had only a gravel surface most of the way, and for many a mile a very different one at that. To my pleasant surprise I now found a first-class motor road extending the whole distance.

On my previous visit I made my way to Ankara with a great measure of uncertainty as to what I should find. I knew to the south of the Taurus I would meet an influence of the Syrian bee and, in the far east, of the Caucasian. But I had no inkling what would await me in the rest of Turkey. Two years previously, at the time I was in Israel, I heard of a book, *Studies on the Honeybee and Beekeeping in Turkey*, by the late Prof. F. S. Bodenheimer who for a short while resided in Ankara. The book was published in 1942. However, it was not until August 1958 that I managed to obtain a copy on loan. A little later Prof. Bodenheimer very kindly presented me a copy. But it was undoubtedly fortunate that I did not see a copy before 1954, for otherwise I might well have written off Asia Minor as of no practical importance in so far as my search was concerned. The book contains many interesting details concerning primitive hives and methods of beekeeping. The chapter on races deals mainly with biometrics and tentative generalisations. The matters of primary importance from my point of view – the physiological characteristics and qualities of economic value – are not discussed. A few are indirectly touched upon, such as colony population counts made in the vicinity of Ankara, but these unfortunately convey the impression that the Central Anatolian bee is the least prolific of any known races, and for all practical purposes of no economic value. Also the statistics cited on the extreme fluctuations in colony numbers in certain areas might well be interpreted as indicating an inherent lack of stamina, or inability to withstand exceptional winters. However, as my findings have shown, none of these assumptions have been borne out by practical experience. But, as I was able to indicate in the preliminary report published in 1958, the Central Anatolian race is of surpassing economic importance.

The first trip to Asia Minor embraced the country between Ankara, Sivas, Erzincan, Bayburt, Trebizond, Samsun, Sinop and Kastamonu; and westward as far south as Eskishir and Bursa; broadly speaking the northern half of Turkey. The journey in 1962 covered the southern half, including the more important sections explored in 1954, but excluding the eastern military zone. The exclusion of this eastern zone was in many respects a great pity, but looking back, it may have been fortunate. Road conditions in Turkey, particularly in the more remote parts, are unimaginably bad, and the ground I managed to cover would have taxed any driver's endurance. Conditions proved nearly impossible east and north-east of Ankara in 1962: the ground had not yet dried out in May and the danger of getting bogged down, with no help in sight, was always present. Rivers were still in spate, and had to be forded without knowing whether the depth of water was too great to prevent a safe crossing. The memory of the hazards, and of the experiences endured, will haunt me for a long time to come. Great improvements in road conditions are now in progress, including the construction of arterial motor roads.

According to the Encyclopaedia Britannica, Asia Minor includes Turkey proper, Armenia, Cyprus and the whole of the Arabian Peninsula. However, my search was on this occasion confined to what is commonly accepted as Asia Minor, the area bounded by the frontiers of modern Turkey east of the Bosphorus and Dardanelles, covering approximately 300,000 square miles, 900 miles from east to west and 300 miles from north to south. This is not a very large expanse of country, but a number of distinctive races have their habitat within it. This may seem surprising without a knowledge of the topography and climatic variations of the area.

Anatolia is ringed by mountain ranges to the north, east and south, and a minor range between the western spurs of the Pontic chain and the Lycian Taurus completes the ring. This western range, though attaining heights close to 8,000 ft. here and there, declines towards the Aegean and Sea of Marmora. In the far east the reverse is true, the highest altitude of 16,916 ft. being attained in Mount Ararat – the traditional resting place of Noah's ark. Within this ring of mountains is Central Anatolia, a steppe about 3,000 ft. above sea level.

Along the coast from Alexandretta to the Dardanelles the climate is Mediterranean, with rainy winters and dry summers. The northern seaboard, from the Bosphorus to Batum, has a heavy

Eydin – An apiary of exemplary orderliness with hives and a size of frame not seen anywhere else.

A comb of Anatolian bees.

A bee-shed on a height overlooking Cankiri, a town north of Ankara

PLATE XIV

ASIA MINOR

Zara, southern Turkey: An outsize bee-bole accommodating a whole stack of cylindrical hives.

Cone-shaped wickerwork hives in the neighbourhood of Isparta, south-western Turkey.

year-round rainfall, increasing as the Caucasus is approached. The average preciptation near the Soviet frontier is about 100 in. (250 cm.). I have a vivid memory of the night of my arrival in Trebizond at the end of August 1954, when it was raining with the intensity we are accustomed to in South Devon. In Eastern Turkey, in former Armenia, the rainfall is by no means as heavy, but winters are severe and protracted. When I was in Erzincan in August 1954 I saw snow from the previous winter still lingering on the surrounding heights.

Central Anatolia has hot and dry summers and severe winters, with temperatures −46°F at Ankara. The rainfall is scanty, averaging 13 in. a year or less. A year-round rainfall, experienced along the Black Sea coast, is unknown in Central Anatolia, where the little that does fall comes mainly in winter and spring. Throughout the greater part of summer this section of Asia Minor presents a spectacle not greatly different from Arabian Desert several hundred miles to the south-east. The huge salt lake, Tuz Gölü, in the heart of this plateau seems only to emphasise the barrenness of Central Anatolia.

In the fertile, semi-tropical plains and sheltered valleys of Cilicia and Antalya, eucalyptus, orange, lemon, date palm and cotton are some of the main sources of nectar. In the rich pastures of the southern slopes of the Taurus, various kinds of clover can be found. In the higher regions, oak and fir furnish honeydew and the alpine flora nectar. On the Black Sea coast we meet a far more varied and luxuriant vegetation than along the Mediterranean, due to the much higher and year-round rainfall, though west of the promontory of Sinop the vegetation tends to be poorer and less varied, as the rainfall gets less towards the Bosphorus. Almost immediately east of Sinop, between Gerze and Alcam, there is a large expanse of jungle, with a richness of vegetation I have not seen elsewhere on my travels. The world's best tobacco comes from the area between Bafra and Samsun. East of Samsun olive and citrus can be seen everywhere; and east of Trebizond tea is extensively grown. On the high ground behind the coastal lowlands are forests of pine, fir, cedar, oak and beech. On the slopes facing north various Ericaceae are common, among them *E. arborea* and ling. Here also are *Rhododedron ponticum* and

R. luteum, from which the poisonous honey is derived.

The vegetation of Western Anatolia is more similar to that of Southern Europe. The area south-west of Izmit is one of the world's most bountiful fruit-growing regions. Though it is known principally for its figs and raisins, fruits of many kinds seem to grow to perfection here. This region is also the most favourable for beekeeping in Asia Minor. Central Anatolia is, on the other hand, the least favourable; spring makes its entry suddenly, with an ephemeral burst of growth, which by midsummer has wilted, the countryside becoming barren, brown and seared. There are almost no trees in this part of Turkey, except round human habitations. The villages and towns of this upland steppe resemble an oasis in summer, but in place of the palm tree the stately poplar reaches skywards. As one would expect, the honey flow in this region is brief but abundant, followed by three to four months of heat, drought and dearth before winter returns. In the spring flush of verdure, many flowers unknown to me graced the countryside. However, judging from the honey obtained and the vegetation I saw, I conclude that the primary sources of nectar are various species of thistle.

To the east of the central plateau, towards the Armenian highlands, the ground rises steadily with a corresponding increase in rainfall and in the severity of the climate. There is a gradual transition in the vegetation, too: beyond Sivas green pastures can be found even in late summer. The honey here is similar to that derived in England from white clover, except that the density is higher.

At Baiburt, at 5,000 ft., the vegetation seems poor and spare; nevertheless I came across some modern hives with two Langstroth supers solid with honey. Kars, close to the Soviet frontier, is reputed to be one of the best honey-producing vilajets; but here, as in many other areas with extensive forests, honeydew is the main source.

Asia Minor has since ancient times been known for its poisonous honey, derived from the violet-flowering *Rhododendron ponticum* and yellow azalea, more correctly *R. luteum*. These two shrubs grow wild en masse only on the Black Sea coast of Turkey, which is their native habitat. The general symptoms of poisoning are nausea, dizziness, headache, blurring of vision and temporary blindness, the severity depending on individual susceptibility and on the amount of poison ingested. Losses of bees have recently been reported in parts of Scotland where rhododendron are grown extensively, but on my visits to the Black Sea coast of Turkey I have never heard of loss of bees from this source.

At the Beekeeping Institute at Ankara I was shown a list of the nectar-bearing flora of Turkey. Included in this list were such well-known sources as lime, acacia and chestnut trees, which I observed now and again, but never in sufficient numbers to constitute a source of any importance. To both my guides beekeeping was an unfamiliar subject, and the language difficulty proved an additional handicap. However, from the information gathered, I was left in no doubt that the diversity of flora offers great possibilities to the beekeeper in Asia Minor.

Agriculture is the main source of revenue in Turkey and the main occupation of most of its people. Great strides have been made in raising the standard of every branch of agriculture since the end of the Ottoman Empire. Every vilayet now has a Director of Agriculture, and in many there is a College of Agriculture where both boys and girls are given free tuition. Beekeeping is included, and in the grounds of these colleges I found almost invariably a large modern apiary; one also had equipment for making comb foundation. There are also experimental centres and nurseries throughout the country, from which the enterprising farmer, fruit-grower or poultry-keeper is supplied high-class stock. At most of these centres beekeeping is represented too, but the most important centre for all matters relating to apiculture is the Bee Institute already mentioned: Türkiye Aracilik Enstitüsü, Umam Müdürlügü, Ankara. A station for raising queens had been established here since my visit in 1954, and as far as I know it is the only place in Turkey where queen rearing is carried out on modern lines.

The Ministry of Agriculture periodically issues statistics which include the numbers of colonies kept in modern and primitive hives in each vilayet, but the figures given cannot be very exact. Large fluctuations in the numbers of colonies often occur, due to drought in Central Anatolia or to other exeptionally adverse conditions in the eastern sections of the country. It is generally assumed that the average number of colonies exceeds one million, of which most are in primitive hives at present.

In no country I visited have I observed such a variety of primitive hives. In the northern half of Turkey, or wherever timber is abundant, oblong wooden hives about 3 ft. × 10 in. × 8 in. (1 m. × 25 cm. × 20 cm.) are in common use. They have a removable cover at the back, or more often a detachable section on top, for taking the honey at the end of the season. Log hives are in use, also logs split in half and hollowed out with a chisel. To get at the honey, the upper half is lifted off with the combs attached. Cylindrical hives in wickerwork appear more common in the southern parts of Asia Minor, but I have seen them here and there in the northern sections. All these hives, with few exceptions, were used in a horizontal position. Occasionally I saw box hives in open sheds stacked in tiers one on top of the other, but more usually they are set out singly. Near Isparta I came across wickerwork hives of about the size and shape of our skeps, but pointed and externally covered with clay. Many odd patterns can also be found on occasion. Clay pipes, used generally in Syria and the other Arab countries and in Cyprus, appear to be uncommon in Asia Minor.

Of modern hives, the Langstroth size and pattern are used almost exclusively, though at Aydin I came across an apiary with hives of an unusual size, fitted with 12 frames about 10 × 10 in., set parallel to the entrance. The hives were expertly made and well kept, and everything indicated that the owner was a keen beekeeper. Near Trebizond I found to my amazement one of the latest fads, a hive fitted with tapering frames as advocated by a French inventor about 15 years ago. Rather surprisingly, at a number of the Agricultural Colleges the hives were of an English pattern with gabled roofs, lifts with splints, porch and alighting board, and legs. How this pattern found its way to Asia Minor I was unable to discover.

The modern hive has not taken on in Turkey as quickly as in many other parts of the world, although the Ministry of Agriculture has spared no effort to get it adopted everywhere.

Apparently the authorities did not at first appreciate that a modern hive is of no value without comb foundation and a honey extractor. On my first visit I observed a great deal of modern equipment in a derelict condition. Where it was in use, I was often confronted by a hopeless tangle of combs, built in any way the bees fancied. One beekeeper, appreciating the need for foundation, fitted the frames with plain sheets of wax, seemingly formed by pouring liquid wax on a slab of stone. It was not surprising therefore that a reversion to the primitive hive took place, for the old-time beekeepers knew how to deal with a colony in it, and how it take the honey at the end of the season. However, on my last visit I was pleased to note everywhere that modern hives were fitted with comb foundation. Great progress during the intervening eight years was apparent on all sides.

ASIA MINOR – 1972

As indicated in the previous report, I visited Asia Minor on two previous occasions: first in August and September 1954 and again in May 1962. These two journeys covered to some extent the same territories, but in the main totally different parts of Asia Minor. In 1972 the areas north-east and north of Ankara were revisited, but some regions adjoining the Aegean in the west of Turkey, which were previously military zones and therefore inaccessible to foreigners, could now be included in our search.

We set out on June 1st from South Germany and in due course passed through Austria, Yugoslavia and Bulgaria. Turkey was reached late in the evening of June 3rd and we spent the night in Edirne. On the way to Istanbul next morning we were expected by Dr. Ismit Imri, who owns an ultra-modern farm and beekeeping establishment north of Silivri. As we were conducted around the farm and apiary we quickly realised that the term ultra-modern proved no exaggeration in any way. Breeding stock was sent from this centre to all parts of Turkey. Through the kindness of Dr. Imri we obtained here our first queens of the Tracian honeybee.

In the course of the afternoon we reached Istanbul, where we were given a warm welcome by Mr. Aktuna, the owner of 'SEDEF'. This firm was founded in 1952 and operates the largest beekeeping appliance and comb foundation factory in Turkey, as well as a modern honey bottling plant. It also runs an extensive beekeeping establishment in the Province of Kars bordering on Russia. At the time of our visit the factory in Istanbul was working to its maximum capacity and the comb foundation mills were in operation day and night. As we could observe, every kind of modern beekeeping appliance was offered for sale here.

Mr. Aktuna and his son offered us their assistance next day. In their company we crossed the Bosporus, to enable us to visit a number of beekeepers on the Asian side – the Bithynia of the Romans. However, we were soon able to ascertain that the indigenous honeybee of this part of Turkey possessed no special characteristics of value. We, nevertheless, obtained some queens from this area for use in our comparative tests.

On June 6th we made our way to Ankara. Dr. Fuad Balci, technical adviser to the Ministry of Agriculture, had been appointed to assist us on the further search in Turkey. Dr. Balci was no stranger to me. Whilst stationed some years previously at the University of Erzerum he kindly procured me some queens of the Armenian variety. On the day following our arrival and subsequent to the completion of all preparatory formalities and tasks, we set to work immediately. A number of beekeepers within reach of Ankara were visited, first one with modern hives, then another owning about 100 colonies in primitive hives constructed of wickerwork. In this instance, to secure the queens, 'driving' had to be resorted to, accompanied by all the romantic aspects associated with old-time beekeeping. Next day we proceeded to Chorum, Samsum and the shores of the Black Sea. The day following we reached Sinop, the most northerly point of Turkey. As our cross-breeding experiments have demonstrated, this seems the area where one of the best strains found in Asia Minor has its habitat. From Sinop our search led due south across the Pontic mountains and as far as Ilgaz, then west to Cherkes, and from there south and back to Ankara, from where the queens we had collected were despatched per air mail to England. Based on the findings we made up to now, the area north and north-east of Ankara appears to harbour the most valuable strains of the Anatolian bee, both from the economic and cross-breeding point of view.

Our further quest on leaving Ankara took us in a south-westerly direction to Burdur. The route to Burdur leads across the central plateau of Asia Minor and past Chay-Afyon and Lake Egridir, where by the waters' edge my previous journey came close to an untimely end, due to a burst tyre. At Chay-Afyon, an unusually keen beekeeper, a teacher by profession, awaited us. In typical oriental courtesy he invited us to stay the night. However, we could not accept the proferred hospitality for we were compelled to keep to our timetable. Shortly on leaving Chay-Afyon a mountain pass had to be negotiated. However, all that afternoon violent thunderstorms raged over this pass, causing near impossible conditions for the crossing, closely bringing our search again to an untimely end. But thanks to the skill and steady nerves of our driver we reached Burdur, though very late at night. Burdur is situated at an elevation of 870 m. by the side of a lake. We therefore found the refreshing air a most welcome change to the heat we had to endure whilst crossing the central plateau. On the other hand, the heat experienced that day was merely a foretaste of what was awaiting us during the coming days. We were now on the fringe of the sub-tropical region of South-Western Turkey – the paradise of Asia Minor. Groves of oranges, olives, figs, apricots, peaches and vineyards extended in every direction as far as the eye could see. The wild flora, where observable, vied in luxuriance and multiplicity with the fruitfulness of these regions and would alone to see justify a visit.

These parts of Asia Minor are, moreover, of immense historical and cultural interest. They embrace one of the main missionary fields of St. Paul. East of Lake Egridir, the part we passed through the day previously, is situated Pisidian Antioch. Today we would pass the site of ancient Collossae and next day Ephesus. We would, indeed, be following in the steps of St. Paul until our departure from Thessaloniki on June 19th.

Kusadasi, by the shores of the Aegean, was our next stop after leaving the heights around Lake Burdur. The way to Kusadasi brought us to Dinar, Aydin and Solke. At Aydin, Mr. Ahmet Istek, whom I met on a previous visit to this area, awaited us. However, his bees had already been transported to the pine forests along the seashore opposite the Greek island of Samos. We immediately made our way to this area in his company. The tracks through these forests, leading to the sites where the hives were situated, proved more difficult then any we had perforce to negotiate on our travels in Turkey. On the other hand, the sub-tropical flora and scenery were entrancingly beautiful. Notwithstanding the near impossible tracks, no less than 60,000 are ostensibly moved annually into this area in August, to gather a crop of honeydew from the firs. Another instance that no effort and hardship can daunt beekeepers.

Mr. Istek most kindly accompanied us to Kusadasi before taking leave. We reached this charming place by the Aegean tired and exhausted after a day of overwhelming impressions and experiences in tropical temperatures of a kind we had not endured up to then. However, next day an interlude of a less exhausting nature awaited us. After a brief visit to Ephesus our next point of call was the Agricultural Institute at Menemen, situated about 45 km. north-west of Izmir. This Institute is doubtless the largest and most progressive of its kind in Turkey. So also is its beekeeping section. Dr. Alev Settar – a pupil of Prof. F. Ruttner – was at the time of our visit head of this department. We also realised that Dr. Settar was a scientist of exceptional ability and one who also appreciated the practical aspects of beekeeping. Under his direction, regular courses on every aspect of advanced beekeeping and the raising of queens are provided. In addition, the Institute places about a 1,000 queens of proven stock at the disposal of beekeepers annually. According to our findings, the indigenous bee of this region is not so hardy and thrifty as the varieties found north and north-east of Ankara. It is also dark in colour, whereas the typical Anatolian is a smudgy orange.

The next lap of our journey took us via Manisa and Akhisar to Balikesir. Dr. Settar offered to accompany us as far as Akhisar. Here we obtained from a commercial beekeeper an almost unlimited number of queens. We were now approaching the part of Asia Minor widely considered as the most favourable for beekeeping in Turkey. We found that the indigenous bee of this region ressembled more closely in colour and markings the typical Anatolian. To reach the beekeepers in the villages in the Balikesir area we were once more forced to hire a jeep. The Mercedes of Herr Fehrenbach would have been wrecked negotiating the perilous tracks leading to these remote sites where the beekeepers kept their hives.

On conclusion of this last search in the wilds of Western Turkey, we were able to wend our way back towards Europe. However, not by the shortest route, for a number of tasks were yet awaiting us in Northern Greece and Yugoslavia.

Leaving Balikesir we followed the route via Edremit and Troya to Intepe. By the Bay of Edremit we passed Mount Ida and the legendary site where Paris made his fateful decision. A few miles before Intepe we came to the crossroad leading to Troya, which Homer in his Illiad rendered immortal. We were, however, by now too exhausted to devote any of our remaining energy to sightseeing, notwithstanding the fact that Troya was so close at hand. A few hours of relaxation from the grinding exertions endured since setting foot on Asian soil two weeks previously was uppermost in our minds.

Accommodation for the night was found in a motel south of Intepe, situated directly by the seashore and the entrance to the Dardanelles. The evening meal was served by the seashore, a mere few yards from the edge of the water. In the light of the setting sun, beyond the southernmost tip of Gallipoli, an opportunity presented itself for a reflection on the events and experiences since our crossing of the Bosporus a fortnight earlier. None of my companions had been to Asia Minor before, nor to any part of the world where life still progresses as it did a thosand or more years ago. Whilst we had still to put up with extreme discomfort and primitive conditions now and again, we were no longer compelled to face the hazards endured on my two previous journeys in Asia Minor.

If one compares the conditions in Turkey today and those of 20 years before, immense improvements are visible in every direction. So, too, in beekeeping, particularly in the rapid change-over from primitive to modern hives. However, less as yet in the adoption of more up-to-date methods of management.

I have in this survey not specially touched on the diverse distinctive races of the honeybee found in Asia Minor, for this has already been done in the earlier report. However, it must be emphasised anew, that Asia Minor represents an unique centre of a series of distinctive races unlike – as far as is known – any other part of the world. In addition, it is the home of a number of races of exceptional economic value and outstanding suitability for crossbreeding purposes. This further search, which covered sections of Turkey which had not been previously explored, confirmed an impression I already gained on my first visit to Asia Minor in 1954.

GREECE

At dawn of June 16th we made our way to the ferry which would take us across the Dardanelles from Canakkale to Esceabet on the European side. From Esceabet we had to traverse the whole length of Gallipoli to reach Kesan, from where the last batch of Anatolian queens was sent off by air mail. The distance from Kesan to the Greek frontier is a mere 30 km. However, our next point of call was Thessaloniki, which could not be reached until next day. From Thessaloniki our search took us west of Edessa, to the Albanian frontier, to secure some queens from this remote section of Macedonia.

SLOVENIA

On June 19th we set out from Thessaloniki for Skopje, Nish, Belgrade, Zagreb and Ljubljana. In Ljubljana, Prof. Ed. Senegacnik and officials of the Slovenian Beekeepers Association, were expecting our arrival. Slovenia is the home of the Carniolan bee and a country of unending interest, and the spirit of hospitality and friendliness of its beekeepers seems boundless. Scenically, with the Karawanken to the north and the Julian Alps to the west, Slovenia possesses a charm and facination all of its own.

Our first task was a visit to the world renowned breeder of Carniolan queens – Mr. Alojz Bukovsek of Medvode – whom I have known for a great many years. Next on the agenda was a visit to the Alpine mating station operated by the Slovenian Beekeepers Associatioon and named after Anton Jansa, the beemaster of Empress Theresia. This mating station is situated close to the limit trees will grow and tucked away in a field of the Krawanken. A chalet, with cooking and sleeping facilities for the caretaker and guests permit a stay of longer duration. A spring of crystal clear water a few yards from the chalet provides an additional need. Queens from this station are sent to all parts of the world and, as one would expect, the demand far exceeds the supply. The track leading to this unique mating station reminded us of the difficulties we had to surmount in the remote Pontic Alps – a further proof that beekeepers will never be deterred by any hardship in pursuit of an objective they have set themselves.

MOROCCO – SAHARA 1976

On my first visit to Morocco, at the end of March 1962, my search covered only a small number of oases where *Apis mellifica saharensis* can be found. Our subsequent cross-breeding and comparative tests were therefore inevitably on too restricted a basis. Nevertheless, the results secured indicated that this race, when suitably crossed, possessed unique potentialities. A further and more extensive search, to ensure the most comprehensive results and benefits possible from this race, was therefore considered imperative. Moreover, from the scientific and practical point of view, we firmly intended to include in this search all the oases where this race is found. As far as is known the range of distribution extends from Ein Sefra and Figuig in the east to the shores of the Atlantic in the west, south of the Atlas. Apart from Ein Sefra and, possibly, Colomb Bechar, both of which are in Western Algeria adjoining Morocco, this race is restricted exclusively to the oases within the bounds of Morocco.

The initial preparations for this further search were set afoot almost immediately on concluding the one to Asia Minor in 1972. As already indicated, we intended to visit every oasis where this bee could be found, to determine possible differences in the characteristics of the bees of one oasis and that of another. The various oases are usually isolated from each other by miles of barren steppe, rocky hills, or endless dunes of sand. The exceptions are the wadies of the Dra and Ziz, where beside the river beds, isolated groups of palms can be observed. These rivers carry water only in springtime and lose themselves in the desert sands.

The date of our departure was fixed on April 20th 1976. But after we had concluded all our preparations, a conflict arose suddenly between Morocco and Algeria over the possession of the former Spanish Sahara. We were informed that all the Passes across the Atlas, as well as all the oases, were occupied by the military. In the circumstances, any likelihood of our being able to carry out our programme seemed out of the question. However, after the lapse of a few weeks, the political tension and risk of a military confrontation subsided somewhat. This gave us hope that we might attempt the journey – though possibly not to the full extent as originally intended. But on reaching Southern Spain further direful reports on the conditions in Morocco placed our attempts in question anew. Very fortunately we did not allow these to deter us from our resolution, and on our arrival in Morocco on April 25th, we found to our relief that the reports in question no longer held good. Indeed, apart from the police checkpoints between some of the oases, we were not seriously hindered at any time during our search in the Sahara.

Herr Fehrenbach again proved our main pillar of support. Mr. and Mrs. Haccour rendered us an invaluable service as interpreters and in the dealings with the native population. Dr. J. F. Corr did so, likewise, in the medical and beekeeping sphere. Mr. and Mrs. Hoenmann, who joined us in Casablanca, volunteered to help in any way possible to them. For a trip of this kind, fully dependable vehicles were of course a paramount necessity. We set out for the Sahara with two Mercedes cars and a heavy Ford van and therefore felt we could confidently face any contingency and danger 'Allah' might decree.

In the Sahara on the fringe of the endless sand dunes.

Zagorra: the niche from which we secured our first Sahariensis queen in 1976.

PLATE XV

SAHARA

Tafilalet: a colony of the Sahariensis in a niche of a garden enclosure wall of clay; the entrance is on the right.

The niche is resealed and the cover boards fixed in place with clay.

We crossed the Atlas on April 28th from Marrakec via the Col du Tichka (2,270 m.). At Quarzazate a brief stop was made before proceeding on our way to Zagora a 171 km. further south. From Quarzazate onwards, the route at first led through a range of barren hills and then along the Dra, which south of Zagora loses itself in the sand. Beside the river bed isolated clusters of palms broke the monotony of the desert scenery until Zagora came in view. Immediately on our arrival at this oasis, Mr. Haccour set out in quest of beekeepers. At supper he informed us that two had offered to part with a queen. However, by next morning one retracted his promise. The other kept his word. He, indeed, proved himself most generous by parting with the queen of his only colony. On completion of our task he furthermore entertained us in true Mohammedan courtesy and hospitality.

At every oasis visited numerous niches for accommodating colonies of bees could be observed in the mud walls enclosing the gardens, but most of them were unoccupied. The fierce climatic conditions and extreme scarcity of nectar sources, apart from the ever-present flocks of bee-eaters *(Merops supercilliosus)*, render the survival of a colony for any length of time well-nigh impossible. In fact, we came across only one garden where a number of niches were occupied. On closer examination some were hopelessly queen-less and doomed to perish. The queens were in all likelihood snapped up by bee-eaters when on their mating flight. Hence the almost insuperable problem of securing a queen of this race at any time.

To find a queen in one of these cavities entails additional difficulties. Every comb has to be cut from the roof of the niche, and due to the extreme nervousness of this race, the queen has in every case to be spotted and caught amongst the milling crowd of bees in the semi-darkness of a cavity, which we found was no easy matter. In a modern hive, particularly with defective combs, the search can prove an even more hopeless task – as we found to our dismay in 1962 when the Governor of Goulmina offered us a queen in a hive of this kind. The combined efforts of three persons were in this case of no avail.

Apart from the unusual nervousness this race, when pure, can be described as good tempered. Indeed, we were able to pull colonies apart without the use of smoke or any other protective safeguards. Mr. Hoenmann, who up to then had only kept the Intermissa, could hardly bring himself to believe there could be so gentle a honeybee as the Sahariensis. On the other hand, as experience has shown, in cool weather or particularly when unsuitably crossed, this race can evince an aggressiveness which hardly bears comparison.

After finishing our task at Zagora, where the heat proved already most exhausting, we returned to Quarzazate and from there turned eastward to Tinerhir, Goulmina and Ksar es Souk. We intended to spend the night at Goulmina to permit us next day to carry out our search there before going on to Ksar es Souk. We were assured accommodation would be available at Goulmina, but this information proved incorrect. We were therefore forced to proceed to the next oasis and to turn twice to Goulmina and, as ill-luck would have it, to no purpose. We were offered a queen, but due to a chain of unfortunate coincidences, I had to depart a second time from this oasis empty-handed. From Ksar es Souk we were able to reach all the further oases we meant to visit. We first called at one south-west of this centre; next one situated to the east and finally Erfoud, 99 km. due south of Ksar es Souk. At Eroud we were received by a beekeeper whom I first met in 1962 and who again most generously rendered us every possible assistance. According to our original plan we would have proceeded from Ksar es Souk eastward to Boudenib and Figuig. The latter is the most easterly oasis of Morocco, in fact on the actual frontier of Algeria. However, in face of the precarious military situation, we refrained from carrying out this part of our programme.

We had by now attained the main objective of our endeavours and therefore ready to return northwards on May 1st – this time via the Col du Zad. But during the day previously violent thunderstorms swept over this section of the Atlas and in the evening before our departure we were informed that the Pass was closed, due to landslides caused by the torrential downpours. This meant that we would have to return to Quarzazate and Marrakech, entailing a hugh detour and loss of time. However, in the course of the night the Col du Zad was rendered passable. So notwithstanding the further omninous signs of thunderstorms, we decided to take the shorter route. This gave us at the same time an opportunity to visit Gourrama, a village tucked away in the southern foothills of the Atlas, situated about 45 km. east of our main route. We were given the name of the beekeeper and a positive assurance we would find the pure Sahariensis in this village. This information, unfortunately, proved false in every respect. All the while we observed further thunderstorms gathering over the heights of the Atlas and our journey was getting progressively more perilous. The risks and dangers did not so much consist in the torrential cloudbursts, but in the floodwaters they created, which rushed with great force across sections of the road often to a depth of 40 cm. The boulders carried along could not be seen and formed the actual menace to our cars. On the northern side of the Atlas snowstorms raged. But, notwithstanding the day's problems and risks, we reached our destination, Fez, safely by nightfall. Fez is widely regarded as one of the outstanding centres of Islamic tradition and culture. To us it was primarily a resort for a much needed night's rest.

A few months prior to our departure, Prof. Ruttner drew my attention to a discovery of a local variety of *A. mellifera intermissa* he made by chance in the Rif of Northern Morocco. According to his biometric findings this variety manifested the largest body, tongue and wing measurements of any known honeybee. He therefore named it *A. mellifera major nova*. In 1962, when exploring the region due west of the Rif mountains, I observed a considerable variation in the colour, behaviour and other characteristics in the Intermissa of that section of Morocco. Seemingly this area of Morocco embraces a centre of variation of the typical jet black Intermissa. Prof. Ruttner suggested we should visit the Rif to carry out an additional survey. This was in fact the primary reason which prompted the diversion to Fez, from where we proceeded next day to Ketama, a small village in the heart of the Rif. On our arrival in Ketama we found ourselves in a different world. A dense fog reduced visibility to a few yards, but next morning a brilliant sun lit up a marvellous mountain scenery with ancient cedars and snow almost within reach.

Our first point of call was Targuist, the place where the legendary freedom fighter, Abd el-Krim, was finally brought to bay by the French and Spanish forces in 1926. A few miles from this town a German Agriculture Organisation established a centre to assist the native population in the development of their land. I was in touch with this organisation before leaving for Morocco and they suggested that our search be extended in two opposite directions, namely, south of Targuist down to the sea and north to the more remote parts of the Rif. As we had only one day at our disposal, we were left with no other alternative but to form two groups. One went in company of Mr. Haccour and an agricultural official from Targuist to Torres de Alcala by the sea. Here tropical vegetation of every kind flourished in great profusion. Thanks to the good offices of the Kaid of this town a beekeeper was speedily found, who offered us four colonies in primitive cork hives. As we had to transport these cork hives in old jute sacks to Targuist we were only able to accept two. The other group, that explored the remote interior of the Rif, where the climatic conditions were the reverse of those by the sea, were equally successful in their quest.

As far as we could determine, the findings of Prof. Ruttner were fully vindicated. Moreover, we found these bees by no means as aggressive as the typical Intermissa of other parts of North Africa. On the other hand, as we subsequently found, the aggressiveness of the Intermissa will show up when crossed including all the undesirable traits of the prototype. Our comparative tests, in addition, demonstrated that the pure and crossed stock manifested an extraordinary high consumption of stores in winter. Indeed, their consumption was more than double that of an Anatolian F-1 in identical circumstances. This wanton extravagance is probably due to an inherent restlessness during the winter months. Whether this sub-variety of the Intermissa embodies any special characteristics of economic importance only time will determine.

On May 4th were were able to re-cross the Straits of Gibraltar, well pleased with the results our search achieved. On the other hand, we could not rid our minds of a fear regarding the future of *A. mellifera sahariensis*. This race managed up to now to survive in the face of every possible hazard and adverse climatic factors, cut off completely and without any contact with the rest of the world since the conclusion of the last Ice Age. However, the random use of insecticides in these oases may well entail its final demise. The permanent loss of this race would unquestionably mean a major loss to apiculture and more particularly in the sphere of cross-breeding.

GREECE – 1977

Already in 1952, during my first visit to Northern Greece, I realised that a time would come when the typical Macedonian sub-variety of the Greek honeybee would disappear. Since the last war modern means of transport have enabled beekeepers in every part of Greece to move their bees to Macedonia annually, to the pine forest of Chalkidiki and the Island of Thasos, to gather a rich crop of honeydew in August. This indiscriminate mingling of genetic dispositions had led to a gradual deterioration of the erstwhile distinctive characteristics of the Macedonia sub-variety of the Cecropia. Indeed, this progressive deterioration during the past 30 years was confirmed by the queens we procured at various times since 1952. We therefore assumed that our only chance of finding the original Macedonian variety would be in the more remote sections of Kassandra, Longos and Athos peninsulas. Our search in 1977 was therefore concentrated on Kassandra and Longos. Athos could not be included at that time, as we did not as yet hold the special permits required for a search on this peninsula. Apart from these two peninsulas, we included the most southerly point of Attica, adjoining Cape Sounion. When I called there previously I was greatly impressed by the particular strain of the Cecropia of this secluded part of Attica, but failed to secure any queens at the time.

As would happen, the results of our efforts in 1977 were severely hampered by unforeseen problems. Greece suffered at the time of our visit from a heat-wave as had not been known in that part of the world for more than 50 years. This extreme heat created difficulties in keeping queens and their escorts alive in confinement. I was well aware that temperatures beyond 95°F could quickly prove fatal. In one instance, to save a valuable queen whose escorts had perished within a few hours of confinement, we took recourse to the expedient of collecting a second set of escorts off the flowers of a shrub by the roadside between Athens and Thessaloniki. This desperate measure saved the queen.

Whilst in some respects the results secured on this journey exceeded our expectations, the tests carried out subsequently on the stock collected clearly indicated that the deterioration we feared had gone much further than anticipated. Whether in the seclusion of Athos, the unique Monastic Republic, we shall find the original Macedonian variety of the Cecropia, remains to be seen. It would, indeed, be a great loss if this particular variety had disappeared in the welter of migratory beekeeping.

THE HONEYBEES OF ASIA MINOR

As we have seen, the Anatolian Peninsula presents every type of topographical variation. The climate ranges from sub-tropical to arid upland steppe and near arctic conditions, all within a relatively small compass. In such a wide variety of conditions one would expect a corresponding variation to the indigenous honeybee. This is in fact the case. Whilst we shall have to await the results of the biometric studies, made on the samples I have been able to collect on these journeys, before final classification is possible, I can indicate in general terms the races I have found, and some of their qualities and physiological characteristics.

Up to now there have been no importations of any consequence in Asia Minor. At the Agricultural Institute at Bursa I was told that at one time a number of experimental importations of Italian queens were made, but that the queens of foreign origin produced bad-tempered offspring when mated to native drones. In view of the unsatisfactory results obtained, the importations were discontinued. Moreover, as modern beekeeping is not as yet practised very extensively, it may be assumed that the bees now found have not been affected by cross-breeding and therefore embody the results of environment and of adjustments effected by Nature from time immemorial. Migratory beekeeping, which would have bearing on this question, is not widely practised except in the western sectors adjoining the Aegean, where also the greatest concentration of colonies is found.

In the most southerly point of Turkey, at Antakya – the Antioch of ancient times – the bees do not differ from A. *mellifera* var. *syriaca*. This is also true at Gaziantep. However, at Mersin, although the bees are still extremely agressive, they appear to me larger and more prolific, and by no means identical in external appearances to the pure *syriaca*. These differences have been confirmed in the crosses we have at present in our apiaries. Further to the north-east, at Malatya, the differences (apart from colour) are still more pronounced. This deep orange colour is found as far north as Erzincan, but I am unable to say how far eastwards it extends. It is not found north of the Taurus. At Gümüsane, some 50 miles due north of Erzincan, we come to a pure black bee which seems to me distinct from the Caucasian we already know. It may appear surprising that within such a short distance from Erzincan a race of bees should be found so different in external appearances as well as behaviour.

These two places are, however, separated by a high mountain barrier, impossible for bees to cross. At Beyburt, 50 miles east of Gümüsane, situated at 5,000 ft. on the fringe of an Armenian plateau, I came across what appeared to me a hybrid. On the Black Sea coast, the dark bee extends westward as far as Samsun. The range of distribution east of Trebizond remains undetermined. We have a few first-crosses of this black Pontic race in our apiaries at present, and have found them prolific, good foragers, but given to excessive swarming. This cross is different in many ways from any of the first-cross Caucasians we have up to now tried out.

We have at the present time pure and first-cross queens under test and observation from places ranging from Mersin in the south to Sinop in the north, and from the furthermost eastern sections of Asia Minor to the most westerly parts – including specimens from the European section of Turkey. But these tests have so far extended over one season only and, unfortunately, over a season which proved a complete failure, and followed the severest recorded winter in this part of the world since 1740. Therefore apart from temper, fecundity, swarming tendency, thrift, wintering ability and a few other characteristics, it has not yet been possible to form an estimate of their relative honey-gathering ability. On the other hand, no better opportunity could have presented itself to test the wintering ability of these races and crosses than the arctic winter of 1962/63. With few exceptions the bees of Asia Minor withstood this test supremely well, both the pure and cross-bred stock.

Whilst it has not been possible to assess the economic value of the importations made in 1962, the available evidence suggests that we shall not find a bee superior to that of Central Anatolia. The first importations of this race reached us in 1955, and I have therefore been able to form a fairly reliable estimate of its economic value.

From its first existence the honeybee has been forced to adjust itself to its immediate environment or perish. The indigenous bee of any particular region reflects in its characteristics the qualities needed for survival in that region. There is, perhaps, no more classical example than the native bee of Central Anatolia – *A. mellifera* var. *anatolica*.

I have already given an outline of the exceptional climate in the upland steppe of Central Anatolia; this in turn governs the flora on which the bee depends for its existence. In the Armenian highlands the winters are admittedly more severe and last longer, but the general conditions are not as exacting as in Central Anatolia – or for that matter anywhere else in Asia Minor.

The honeybee of Central Anatolia is of unimpressive appearance. She is small, resembling the Cyprian in size, but has none of the glamour or uniformity in colour of that race. The colour of the Anatolian bee can best be described as a smudgy orange, turning to brown on the posterior dorsal and ventral segments. The scutellum is usually dark orange. The queens have a dark crescent-shaped rim on each dorsal segment – a characteristic of all Eastern races – but here they are browny black, and in place of the yellow or light orange we have a dark orange. But beneath this sombre exterior are hidden qualities of incomparable economic value.

The Anatolian tends to extremes in both its good and bad qualities. Fortunately, she has few undesirable characteristics, the most serious of them being her disposition to build a brace-comb beyond all reason. This is of no great consequence in primitive beekeeping with fixed combs, but an excess of brace-comb renders null and void the essential advantage of a modern hive. The Anatolian, in addition, uses propolis freely, which accentuates the drawbacks of the brace-comb. However, both these defects are largely mitigated, if not eliminated, when queens of this race are crossed with a good strain of Italian or possibly Carniolan. Indeed, it is only when suitably crossed – either in a first- or second-cross – that most beekeepers can hope to secure the best economic results from the Anatolian bee.

As for her good qualities, I believe I can state in all truth, that the Anatolian stands beyond comparison – certainly in foraging powers, thrift and wintering abilities. When crossed, she is extremely prolific. By mid-June a 12-frame Modified Dadant brood chamber will usually be found chock-a-block with brood and honey. However, she does not breed to excess out of season, as so many other races are disposed to do. She is slow in building up in spring; she will not make a determined effort at extending her brood nest before settled weather has set in, but will then outstrip every other race. She does not squander precious stores and energy in premature and futile endeavours, in changeable and unfavourable early spring weather. After the main honey flow, and in times of dearth, she contrives to husband her reserves of stores and energy in an uncanny way. I regard the thrift of the Anatolian – particularly in our uncertain climatic conditions and honey flows – as one of her most valuable economic qualities, a quality which is sadly lacking in so many of our present-day strains, which breed to excess in times of dearth. Experience has shown that the Anatolian bee will take care of herself in times of dearth and in seasons of failure, when others die of starvation.

I have stressed the great fecundity and breeding powers of this race. I would, however, point out that, were it deemed desirable, one could without much difficulty develop a strain by selection which would accommodate itself readily to a single brood-chamber of British Standard dimensions.

Though so prolific when crossed, the Anatolian is not given to swarming, as our experience has demonstrated. She is also very good tempered, bearing manipulation with the greatest calm and composure, although she defintely resents interference in cold weather and late in the evening. Moreover, with regard to temper, there appears to be a considerable variation in strains, as I could verify myself when in Turkey. But the Anatolian is no exception in this respect; there is to my knowledge no race which does not show up a difference in temper between one strain and another. When unsuitably crossed, or when mated at random to drones of unknown origin, bad temper will result in almost any strain or race.

As already indicated, the Anatolian is endowed with an inexhaustible capacity for work – a faculty which enables her to turn her other good qualities into something of concrete value. Indeed, this bee embodies the highest development of industry and honey-gathering ability of any race known to me. In addition, we have here a bee that not only does extremely well in a good season, but one that does exceptionally well in indifferent and poor seasons. This is of far greater consequence and practical importance than a surpassing performance in an occasional really good season. The ability to do well even in the poorest of summers was clearly demonstrated during the disastrous season of 1963. On the other hand, in the exceptionally good season of 1959, when our average honey yield amounted to $169\frac{1}{2}$ lb. per colony, the Anatolian crosses far exceeded this figure, and fulfilled our expectations in every way.

The Anatolian possesses many qualities and characteristics which may bewilder those who are unaquainted with the peculiarities of the race. For instance, Anatolian queens will usually take up to a week longer before commencing to lay after mating. This peculiarity has seemingly nothing to do with the weather, for the same delay would occur when under ideal mating conditions. On the other hand, I have found that 25 per cent of the queens will give a full four years' service, with unimpaired energy and fecundity, even in a normal honey producing colony. It may be assumed that this exceptional longevity – which is most remarkable considering the great fecundity of the queens – is in some measure transmitted to the worker progeny. The extraordinary strength of the colonies, in relation to the actual fecundity of the queens, could not well be explained otherwise.

I wish to emphasise onece more: the pure Anatolian cannot be relied on for maximum performance. It is only when suitably crossed that the full economic potentialities of the race come to the fore. Furthermore, as no selection has been done in its homeland up to now, queens of the best stock are not readily obtainable. But no doubt, in view of the great progress now under way in Turkey, the prospects of obtaining select breeding stock should materially improve.

Whilst I have had the good fortune to discover in the Central Anatolian bee a race of surpassing economic value, the two journeys to Asia Minor were accompanied by untold vicissitudes and difficulties. I was also compelled to cut short the programme in 1962 due to an accident. When travelling beside the shores of Lake Egridir a tyre burst – though I had special heavy duty tyres fitted as a safeguard against such an occurrence. The car plunged down the high embankment and overturned on a heap of rubble. Fortunately, the damage was mainly superficial. With the arrival of help, the car was put back on the road and we were able to proceed to the next village. For the essential repairs I had to wait until I reached Salonika some weeks later.

THE AEGEAN ISLES

After completing my task as best as I could in Asia Minor, I proceeded via Edirne and Kavalla to Salonika, where the repairs to the car necessitated a week's stay. I availed myself of this opportunity to make a further exploration of the Greek section of Macedonia. The authorities at the American Farm Institute kindly provided the necessary facilities.

It was in 1952 that I sent the first consignment of Greek queens to England. With the assistance of the American Farm Institute I was able to procure a further supply from the Chalkidiki Peninsula. The original stock imported in 1952 gave us extremely good results, and the intervening years have in no way diminished my first appreciation of the value of this race. Indeed, I consider this as one of the most valuable races we have. I was therefore very glad of the opportunity to procure a new supply of breed stock.

In 1952, when I explored the mainland of Greece and the Peloponnesus, I included a visit to Crete. It was then already realised that my search would not be complete without exploring some of the Aegean Isles – notwithstanding the many difficulties a visit to these more remote islands would entail. The Aegean comprises 483 islands, and it was not clear at the outset that only a few could be visited. Indeed, many of the islands are barren, or the vegetation so scanty as to preclude the necessary subsistence for bees. On the other hand, there are islands such as Thasos, Icaria and Samos, with an exceptionally high honeybee population. I was able to include these islands in my programme in the autumn of 1954, after my visit to Asia Minor.

The Aegean is in many respects a veritable fairyland, but a voyage to the islands can – except on luxury tourist boats – prove a most unpleasant experience. The small steamers that ply between the islands carrying freight and passengers are often loaded with cattle, domestic animals, fish and human beings to the point of suffocation. When, as is often the case, the treacherous currents in the narrow straits contribute their part to the discomfort of the passengers, the result defies description.

My first objective was the island of Ios, near the centre of a group known as the Cyclades. It was fairly certain that the bees on the other islands would not be substantially different. Another voyage to the Southern Sporades included Samos and Ikaria. The most southerly islands of this group were purposely avoided, because of the likelihood of importations during the Italian occupation. I intended to visit the most northerly island of the Aegean, Thasos, renowned for its honey and beekeeping, but a fortunate circumstance made this unnecessary.

The island of Ios comprises about 46 square miles, and has a population of approximately 7,000. According to tradition Homer was buried there. At the time of my visit the bee population numbered about 3,000 colonies, of which 550 were in modern

Wild thyme growing in a cleft of a rock. Some Aegean islands, particularly Anaphi, are renowned for their thyme honey.

PLATE XVI

AEGEAN ISLANDS

Ios: On the way to the heather – the beekeeper carries two and the donkey four cylindrical hives.

Thasos – known as the "Bee Island" because of its phenomenally large number of colonies.

A primitive beekeeper on Thasos engaged in the old-time way of harvesting the season's crop.

hives. Ios is very mountainous, and all the hives were at the heather, on the heights of the island. As there are no roads, to get to the bees we had to avail ourselves of donkeys or mules, the only means of transport at hand. This was a slow and arduous way of getting about. However, the hives, both modern and primitive, were carried up by the same means. A donkey carries four primitive hives, the beekeeper trudging along on foot with one on his shoulder and another tied to his back. These poor islanders spare no effort, and a more arduous mode of transportation could hardly be visualised.

We had to set out from our hostel by the quayside before daybreak. The party comprised nine persons, and most of the way we had to ride in single file along a treacherous track. As dawn broke I observed at first a great variety of sub-tropical vegetation. Then at higher altitudes I saw more and more heather. Though *Erica verticillata* was most common, I could observe other varieties previously unknown to me; unfortunately, no one in the party could give me their botanical name. Gradually a group of hives came into view here and there, sheltered in a hollow or the lee of a rock, but never more than ten to 20 in one place.

There was no doubt the bees here belonged to the same race as that found on the mainland of Greece. Very strangely I could observe the same phenomenon here as in Crete, namely, an occasional colony displaying a stinging propensity rivalling that of some Oriental races. The majority of colonies were in every way as good tempered as those of the mainland, where I never met an instance of this extreme irritability. Such isolated manifestations of extreme bad temper are difficult to explain, for there were no visible indications that this was the result of an importation from the Near East.

We had to remain on Ios for two days until the arrival of the next boat. We then proceeded to Sikinos, Pholegandros, Santorin – the black pear of the Aegean – and to Anaphi, the most southerly isle of the Cyclades. Anaphi is renowned for its thyme honey, and whilst our boat was at anchor in the roadstead, the evening breeze was laden with the heavy scent of the thyme. On our return voyage to Piraeus we passed Amorgos, Naxos, Mikonos and Syra, where we called on the outward passage and, on this occasion, for a brief interview with the Director of Agriculture in charge of the whole of the Cyclades.

I had only a short stay at Athens until the Ministry of Agriculture completed the necessary arrangements for my visit to Samos. Samos is renowned in many ways, perhaps foremost for its muscatel. It is a highly fertile island of about 180 square miles, with a population of 67,500. There are 4,855 colonies of bees, of which 3,480 are in primitive hives. The next largest island, Ikaria, though only about half the size of Samos, has 8,240 colonies according to the figures submitted to me by the Director of Agriculture at the time I called. Both Samos and Ikaria are under the jurisdiction of the Director at Vathy Samos.

According to these figures the colony density on Ikaria is about 901 hives per square mile, and probably the highest in the world. Thasos, in the northern Aegean, which is about one-third larger, has 10,000 colonies and is commonly referred to as the 'Bee Island'. On both these islands honeydew, derived from a pine, *Pinus halepensis*, constitutes the main crop. However, on Ikaria *Erica verticillata* seems of similar importance. As far as I could ascertain, Ikaria and Thasos, with Chalkikidi – the Peninsula on the northern shores of the Aegean – are the most important bee-keeping centres of Greece, and areas where the production of honey forms the sole means of livelihood of many beekeepers.

The bees of Samos and Ikaria are apparently of Western Anatolian stock. The nearest point of Samos is barely one mile from the shores of Asia Minor, and Samos and Ikaria lie 11 miles apart. When on my way from Aydin to Ephesus in 1962, Samos was clearly visible from the main road some miles inland.

As already indicated, there was no need for me to visit Thasos. When en route from Istanbul to Salonika a few weeks earlier, I had an opportunity to visit Philippi, a few miles off the main road between Kavalla and Salonika. I felt I ought not to miss this chance to visit the site where St. Paul founded the first Christian community on European soil – apart from the other historic associations of Philippi. So when I turned off from the main road to Salonika, my thoughts were in the past. But before reaching Philippi I observed on the plain to the left of the road an enormous array of wickerwork hives, lined up row upon row. The regular layout was clearly the handiwork of a beekeeper who took great pride in his calling. The hives were all of one pattern and of enormous capacity: the apiary was the work of an exceptionally competent beekeeper, owning an unusually prolific bee.

Apart from their great capacity, these hives had another interesting feature. The vertical staves of the wickerwork projected a full 2 in. at the bottom, and therefore permitted the bees' entry and departure in every direction *ad libitum*, and an amount of ventilation far in excess to what is usually considered necessary. This seemed all the more striking, for the beekeepers in Greece usually keep their hives' entrances contracted far more than is customary here in England.

I was informed that these hives came from the isle of Thasos. They were brought here at this time of year, when there was nothing for the bees to gather on the island whereas here on the mainland they were able to eke out an existence. The large number of colonies on one site, their excellent condition and the exceptional capacity of the hives, conveyed to me all the information I required concerning the bees and beekeeping on this island.

From these details it will be appreciated that beekeeping on the islands of the Aegean is a factor of major economic importance. Although the bees on some of the islands may not possess any special value for breeding purposes, their productive and economic value cannot be questioned. No one can make a livelihood with bees of inferior stock – particularly where, as is usual here, primitive beekeeping is the rule rather than the exception.

YUGOSLAVIA: BANAT

It is generally conceded that the most typical forms of *A. mellifera* var. *carnica* are found in Upper Carniola and the two adjoining Provinces of Carninthia and Styria. In the English-speaking countries this race is commonly known as the Carniolan, for the first importations – indeed, most of the importations up to 1940 – came from Upper Carniola. However, the geographical distribution of the race extends far beyond the three Provinces named. As far as we know at present, it extends to the whole of Yugoslavia, Hungary, Rumania, Bulgaria and the greater part of Austria. But precise details are lacking. The Greek bee, *A. mellifer* var. *cecropia, is doubtless a sub-variety of carnica.* In appearance the two races do not differ, but there are clear and decided differences in their physiological characteristics. As far as I have been able to ascertain, the bees of Northern Greece, particularly those of the Chalkidiki Peninsula and the narrow strip of country between the Aegean and Rhodope mountain range, including both Greek and Turkish Thrace, owe their economic superiority to an influence derived from the Anatolian bee. How far the Anatolian influence extends into Bulgaria, into the plain of the Maritsa, we do not know. There are undoubtedly considerable variations the further we get away from the main centres of the habitat of *carnica*. In fact, even within the confines of Yugoslavia, considerable variation is found – though externally the bees differ little or not at all from *carnica* as generally accepted.

I made an extended trip through Bosnia, Herzogovina, Montenegro and South-Western Serbia, and found the bees of these areas more prolific and less given to swarming than the true *carnica*. However, they are more disposed to propolise, and were apparently also more susceptible to nosema. Indeed, this susceptibility was so highly developed that we could do very little with these strains here in England.

References appear now and again in apicultural literature concerning a sub-variety of *carnica* found in the Banat – a region situated where the frontiers of Yugoslavia, Hungary and Rumania meet. This bee attracted attention already more than 100 years ago. However, most of the references I have been able to trace confine themselves to a bare statement of the existence of the race, and precise details of its characteristics and economic value have hitherto eluded me. That this bee of the Banat should have attracted special attention more than 100 years ago seemed to justify further investigation, but no opportunity presented itself until 1962.

The Banat is situated to the south-east of the present Hungarian frontier, enclosed by the Danube to the south, the River Moros on the north, the Theiss on the west and the Transylvanian

Alps on the east. During the Turkish occupation from 1512 to 1718, the land was allowed to go derelict, but under Maria Theresa immigrants from Western Europe were encouraged to settle and re-populate it. Evidence of this re-colonisation are still visible everywhere. However, the Banat is no longer an integral unit; one-third of its area now belongs to Yugoslavia and the rest to Rumania.

I had often heard of the extensive acacia forests of this region, and as I travelled north from Skoplje I observed the acacia in bloom everywhere. Hence on my arrival in Belgrade I was not surprised to find that the hives had been moved east to the acacia woods close to the Rumanian frontier.

The extensive acacia forests are confined to areas of poor sandy soil which could not be utilised to good purpose in any other way. Maria Theresa had them afforested with acacia – one of the few plants that would thrive there. Beekeepers are now the beneficiaries of the undertaking initiated by the Empress. It was at once apparent on our arrival that the honey flow was at an end: with every breath of wind the faded blossoms tumbled off the trees like snow. But the bees had done well. There was an atmosphere of prosperity in the clearings in which the hives were placed.

I was given complete liberty to examine any colony I pleased. As the hives were full of honey, such examinations were not easy, but they were facilitated by the remarkable docility of the bees, which made it possible to perform the work without a veil. Breeding had been severely restricted by the heavy flow, and I could not observe any signs of swarming. One thing struck me at once: the Banat bee showed very much more colour on the first three dorsal segments than I had hitherto observed in any of the *carnica* strains. The colour was not such a clear yellow as found in the Italian, but a tawny yellow or rust-brown usually associated with the primary race. However, in the true *carnica* the rust-brown shows up only now and again, and never to the same extent as in the Banat bee. There is usually a fairly wide variation in the colour of the Banat bee, and in some it might be described as yellow. The scutellum of the workers varies from yellow to brown; the overhair is light brown, and the tomenta grey with a tinge of yellow.

We do not know how this sub-variety originated. As already stated the Banat bee was regarded as a distinct race long before any large-scale interchange of queens took place from one distant region to another. Indeed, the modern hive had only just been invented, and until then an exchange of queens was hardly a practical possibility. The settlers of Maria Theresa's time came from parts of Europe where none but the common black bee was known. This bee seems to be to all intents and purposes a freak of Nature – brought about by a chance combination of factors that form part of the genetic make-up of *carnica*, and which manifest themselves spasmodically in the rust-brown markings that cause so much concern to the present-day breeder who aims at perfect uniformity. The fact that this bee has been able to assert and maintain its distinctiveness, in the heart of the habitat of its parent race, is surely a remarkable phenomenon.

EGYPT

It was a typical cold, damp and misty October morning when I set out on the first stage of my journey to Egypt. But at Zurich it was warm and sunny – one of those charming autumn days so characteristic of this part of Switzerland. By 9.30 that evening the lights of Alexandria became visible: when we landed the temperature was 81°F.

A group of officials from the Ministry of Agriculture and Cairo University, headed by Dr. Salah Rashad, welcomed me at the foot of the steps from the plane, and within a short time I was in Cairo.

Before setting out for Egypt some last-minute difficulties threatened to upset the long-planned arrangements. Prof. A. K. Wafa, who undertook all the preparations, and who had keenly looked forward to my coming, suddenly fell ill. He had to come to London for treatment, and I met him there the day before I left England. However, Dr. Salah Rashad of Cairo University, and Dr. Mohamed Mahmood of the Ministry of Agriculture, jointly took over and carried out the arrangements which Prof. Wafa set on foot. At the conclusion of my visit to Egypt, in fact the day before I left, Prof. Wafa returned to Cairo and was able to join the farewell celebrations organised by Dr. Rashad.

When Egypt is mentioned in apicultural circles, a picture is conjured up of the migratory beekeeping carried out on the Nile in the time of the Pharaohs. From all we know, it may be safely assumed that beekeeping played an important part in the life of the people of the Nile Valley from time immemorial.

We know all too well that wherever beekeeping is practised, success and failure revolve round a nicely balanced adjustment of sunshine, warmth and moisture. In the Nile Valley these essentials are virtually never absent. Admittedly, sunshine and warmth vary in degree, but they can be relied upon as surely as night follows day. The needed moisture is here not dependent on the fitful vagaries of wind and rain, but on the life-giving waters of the Nile. Of winter as we know it, there is none. The nights can be fresh from mid-November to mid-February, but the temperature rarely, if ever, drops to freezing point. Temperatures are high in summer, reaching up to 110°F at Cairo and even higher further south. However, the extreme heat is tempered by an almost constant wind from the north throughout the year, without which the climate would be very trying indeed.

Of the total area comprising Egypt, 96.5 per cent is barren country. Excluding the larger oases, all agriculture and beekeeping is confined to the Nile Valley and Delta. The great triangular area of the Delta, the most fertile region of Egypt, extends 100 miles from south to north and to a width of 155 miles along the sea from Alexandria to Port Said. The rich alluvial soil varies here in depth from 55 to 70 ft. Every square yard of this area is under intensive cultivation, as is also all the arable land within the confines of the Nile Valley. The width of the valley varies from six to 16 miles. Towards the far south the width declines to one or two miles, and on approaching the Sudan all vegetation ceases. All along the Valley there is a sharp line of demarcation on both sides between the cultivated ground and the fringe of the barren regions.

Great efforts are in progress to bring limited sections of the desert under cultivation, particularly within the confines of

Kharga and Dakhla oases, situated far south in the Libyan Desert – now referred to as the New Valley. Smaller schemes are under way in the Wadi Natroun and at Mariout between the Nile Valley and the desert road from Cairo to Alexandria. Areas which a few years ago were destitute of vegetation are now producing crops of many kinds, as I could witness myself. Bees and beekeeping closely follow on the heels of these new ventures.

Unlike other countries, Egypt does not possess a wild flora of value to the honeybee. Nor are there any woods or forests as we know them. Apart from the date palm, eucalyptus and citrus, few trees of any importance to bees are found. The main nectar-bearing sources are cultivated crops of one kind or another. The date palm, the most common tree in Egypt, is of considerable value as a source of nectar and, on occasion, when the fruit is fully ripe, bees will collect from dates an almost black syrup, of which I was shown a sample at the Ministry of Agriculture. Eucalyptus trees border the main roads everywhere. But the orange blossom is doubtless one of the most valuable sources of nectar. In Spain, Greece, Turkey and Palestine the orange groves are confined to certain limited areas, but here they are dotted about the whole of the Delta and Nile Valley, and also in the larger oases.

Of the cultivated crops, Egyptian clover (*Trifolium alexandrinum*) colloquially known as bersim, forms one of the primary sources of nectar. This clover constitutes the main fodder crop for cattle and is therefore grown everywhere. Cotton is another major source, which is again grown everywhere. Indeed, one-third of the total arable land in Egypt is devoted to the cultivation of cotton. According to the information given to me, there is a substantial difference in the amount of nectar secreted between the various varieties of cotton, short-fibre cotton secreting most abundantly. Broad beans (*Vicia faba*) are extensively grown and play an important role in beekeeping – not as a source of surplus but for building up the colonies. Beans commence to flower about mid-December and it is on the nectar derived from this source that colonies build up for the main honey flows. Whilst almond, apricot and peaches provide subsidiary help, apples, pears, plum and cherries are not grown extensively in Egypt; on the other hand, many kinds of sub-tropical fruit develop to perfection here, and no doubt some of these are of limited value to the honeybee. Maize, rice and sugar-cane are grown extensively, but these are of no value as sources of nectar.

In many parts of the world, when a particular nectar-bearing source is under consideration, we automatically qualify its value according to its dependence on the right type of weather. No such qualification is called for in Egypt, except for the period of the khamsin – the hot, dry, sand-laden winds of the spring months coming from the south. When these prevail, the sun is obscured, and their fiery breath can wilt the most promising display of blossom in the matter of a few hours.

I have been unable to ascertain when and by whom modern beekeeping was first introduced to Egypt. Modern hives were apparently found here and there in small numbers at the turn of the century. However, I believe it was the late Dr. A. Z. Abushady, on his return to Egypt in 1926, who – with his habitual drive and enthusiasm for modern developments – initiated the real break with the traditional ways of beekeeping. In a garden of a suburb of Cairo I was shown English hives containing specimens of the aluminium comb foundation invented by Dr. Abushady during his stay in England. At an early stage the Langstroth hive was adopted, and this is now used exclusively, with a few minor modifications here and there.

Great progress can be discerned in Egypt in every sphere of endeavour, including the advancement of modern beekeeping. The Ministry of Agriculture has even been considering the possibility of enforcing the universal use of modern hives throughout the country. But it is realised that the time for such a step has not yet come. Nevertheless, in view of the great difference in the amount of honey obtained, the modern hive will in due course inevitably supplant the old-time methods. I have it on good authority that the yields from modern hives average 60 lb. and from the primitive hives 6 lb. (27.3 kg.).

Throughout the whole of Egypt I saw only one kind of primitive hive, of the sun-baked clay cylindrical pattern. These hives vary a little in size, but are usually about 46 in. long and 8 in. in diameter internally. The walls are about $1\frac{1}{2}$ in. thick, and therefore of substantial construction and weight. However, there is no

The honeybee and lily were the emblems of the Pharaohs. "Lord of the Bee" was the ancient title of the Kings of Lower Egypt.

For all we know sun-baked clay hives have been in use in Egypt from the beginning of time. Stacks of newly-made hives in the foreground.

PLATE XV

EGYPT

Minstry of Agriculture isolation apiary at the oasis of Farjum, reserved for the breeding of Italian queens.

An apiary guard whose all-day task is the destruction of hornets.

South of Asyut: A typical stack of Egyptian hives of which many provide room for more than a thousand colonies.

migratory beekeeping, and they are handled but once when placed in their permanent position in a stack. These hives are never used singly, but always in stacks, tiered one on top of each other, seven to ten high. When in their permanent position, the spaces between the tiers are filled with clay at both ends. A disc of sun-baked clay, fitted to the back and front of each cylinder, completes the individual hive. The disc in front is provided with a small aperture for an entrance, which is at the top, not at the bottom as is commonly the custom in other countries. These stacks of hives convey the appearance of gigantic blocks of clay, only the outline of the individual cylinders showing at the front and back. In the Delta one can often see two or more of these stacks set one behind the other, but with an adequate working space between, and 200 to 300 hives in each stack. On the other hand, depending on the space available, single stacks extending up to 150 ft. in length are not uncommon. In one such stack I counted no less than 1,200 hives. In Upper Egypt smaller stacks are more usual, with 150 to 200 hives.

The silt carried down by the Nile turns extremely hard when baked in the sun. For material for making the primitive hives, the Egyptian beekeeper uses the alluvial soil he finds wherever he may be, and he mixes this with finely cut straw. The hives cost him nothing, apart from the time and effort needed to fashion the cylinders. Moreover, these clay hives, stacked in the way they are, ensure the most efficient protection against the extreme heat and fierce rays of the sun. In Egypt modern apiaries must be provided with an overhead shelter or shade of some other kind. It would be inviting disaster to expose modern hives to direct sunshine.

Special tools are used with these primitive hives, each designed to facilitate a particular operation. An iron tool, resembling a miniature sickle of stout construction, is used for prising open the cover-plates. A number of instruments made of steel, about 4 ft. long and furnished with a hooped handle to give the necessary control and purchase, are also used. One has a large bowl fitted to the far end, and this is used for hiving swarms; the end of another is shaped like a double-edged sword, for cutting free the combs when the honey is taken; a third is fitted with a hook or prongs for withdrawing the honey combs from the cylinders. The interior of the hives is lit up when necessary with the aid of a mirror. To drive the bees into the forward section, and to keep them in subjection while the work is in progress, a whiff of smoke is applied now and again from a smouldering cake of dried camel dung.

For all we know, this method of beekeeping may date back to the time before recorded history. But one thing seems evident: the hives used by the ancient Egyptians on their migratory expeditions up and down the Nile could not have been made of clay. The weight and near impossibility of handling such hives seems to preclude their use in this manner.

Beekeeping has always played an important role in the life of the people of the Nile Valley. That this is so at the present time no one will question. I was in fact greatly surprised at the extent and importance attached to beekeeping in modern Egypt. No precise figures are available as to the number of colonies, but reliable estimates place them at 1,500,000. Apiaries of over 1,000 colonies are common. On the outskirts of Damanhur there is a modern apiary of 400 colonies. In the Province of Tanta there are 21,000 modern hives and 107,000 primitive ones. Several factories specialise in the manufacture of modern hives, and the appliances required by modern beekeepers, including comb foundation. The export of honey is engaging the attention of the Government.

In several respects Egypt ia ahead of many other countries. The over-riding importance of reliable breeding stock is here fully appreciated. Bee breeding stations play a vital role in the advancement of apiculture. A number of them are operated by the Ministry of Agriculture and others by private enterprise. One such station, owned by the Government, is at Fayum near the southern shore of Birket Qaroun. Here pure Italians are bred in complete isolation. Another in the north, on a peninsula jutting into Lake Mansala, west of Port Said, is operated on a co-operative basis and is reserved for breeding Carniolans.

Importations have been made for a great many years. At one time, before 1922, Cyprians were imported, but this race has gone out of favour completely. I believe it was Dr. Abushady who popularised the Carniolan bee. At the present time the great majority of the modern hives are stocked with pure or hybrid

Carniolans. I came across only one commerical beekeeper who favoured the Italian bee. Some of the American strains are now also being tested, but the Carniolan is certainly the most popular bee in Egypt today. Moreover, no efforts are considered too great to ensure pure mating and to secure pure stock of this race for distribution to beekeepers. The native bee is practically nowhere kept in modern hives.

The Ministry of Agriculture operates a number of breeding establishments, as well as that at Fayum: there is one at Borg el Arab, west of Alexandria, and a whole series at the oases in the southern Libyan Desert. It also runs a Quarantine Station at Noubaria, where all imported queens are retained for a certain period. The Ministry has furthermore its own Experimental Stations for research purposes. However, the primary apicultural research is carried out by the Faculties of Agriculture of the various Universities. Foremost among these is Cairo University, headed by Professor A. K. Wafa. At Ein Shams University, Dr. M. A. El-Banby has concentrated on the biometric and biological studies of the Egyptian bee; at the University of Alexandria, Professor El-Deeb is in charge; and at Asyut University, Professor M. H. Hassanein. An extensive experimental apiary is attached to each of these universities.

The Bee Kingdom League – the Association founded by Dr. Abushady – takes a very active part in the advancement of apiculture, and it publishes a bee journal in Arabic giving the latest information of every aspect of beekeeping. I met all the principal members of the League at a reception organised for this purpose.

Throughout my stay in Egypt I was pleasantly surprised at the keenness evinced by the authorities, as well as by the individual beekeepers, to gather any information they could. There was everywhere a great eagerness for advice and particulars of the latest advancements in all apicultural fields. To meet this demand, I gave a series of lectures on the more advanced aspects of bee breeding. In Egypt the main efforts have hitherto been concentrated on securing purity of stock. I suggested that the work should be carried a step further, to the improvement of strain by selective mating in isolation. Egypt is in an ideal position to lead the world in this. The numerous oases, guaranteeing absolute isolation, together with such favourable climatic conditions, give all the essential requirements for an intensive scheme of selective breeding of the honeybee.

THE EGYPTIAN BEE

The indigenous honeybee of Egypt, A. mellifera var. fasciata, has aroused interest from the very inception of modern beekeeping. As early as 1864 importations were made into Central Europe for the purpose of ascertaining its economic value. Further importations were made early this century, though purely for scientific purposes, by Professor H. V. Buttel-Reepen, and also by Dr. Egon Rotter, who at that time lived in Czechoslovakia. Whilst in this country Dr. Abushady made an effort to promote the importation of the Egyptian bee, but he did not meet with much success.

The Egyptian bee is unquestionably one of the most intriguing honeybee races. It is the smallest honeybee – apart from Apis florea. In its native habitat I observed here and there individual bees not much larger than our common house fly. But its appearance would capitivate the imagination of every lover of the honeybee. The bright orange colour, and particularly the nearly white pubescence – which makes the bee appear to have been dusted in flour – gives it an irresistible charm. The bright orange extends to the fourth dorsal segment; the ventral segments are almost completely yellow, excepting the last two which are dark. The thorax, and the dark colouring of the dorsal segments, are jet black. The scutellum of the workers is bright orange, but that of the queens and drones is black.

As would be expected, the queens are smaller than in any other race. The abdomen of the queen is a bright orange, with a narrow sharply defined crescent-shaped rim to each segment – the characteristic marking of all oriental races. I have not yet been able to assess the fecundity of the queens; according to to Dr. El-Banby they are not prolific in comparison with the other races, and this is doubtless so.

The pure *fasciata* is according to all accounts greatly addicted to swarming; this must be a hereditary disposition, for the primitive hives are fairly capacious and do not restrict breeding. Many swarm queen cells are constructed; they are usually not built singly but in clusters even on the face of the combs – a characteristic I have not observed in any other race. Anatolians, Syrians and Cyprians will construct queen cells in clusters but always on the edges of the comb. The queen cells of *fasciata* are small and almost smooth.

The natural comb of the Egyptian bee has smaller cells (32-33 per square in. instead of 28), but I find brood rearing progresses normally in comb with cells of common size. The honey cappings are exceedingly watery, far more so than those of any other race. However, this bee does not propolise – a rare quality which she shares with the Indian races. For breeding purposes, I consider this as one of her most valuable qualities. (It must not be assumed that there is no propolis in Egypt: at Fayum, where Italian queens are bred, I found the interior of the hives plastered with the most resinous type of propolis found anywhere.) Another of the valuable qualities of the Egyptian bee is her highly developed instinct of self-defence and disinclination to drift. The two qualities are complementary, and with the close stacking of the primitive hives, with no distinguishing marks between one entrance and another, drifting and lack of self-defence would create an impossible situation.

I heard various opinions on the honey-gathering ability of the Egyptian bee. In its native environment it must be fairly productive, judged according to the capacity of the primitive hives and the relative fecundity of the queens. One of its greatest drawbacks is undoubtedly its temper. However, in some sections of the Delta I found reasonably good-tempered bees. In other parts, particularly in Upper Egypt, the reverse was the case.

Most remarkable of all, the pure *fasciata* has no ability to form a winter cluster with the onset of severe cold, as experienced in temperate zones. She may never have possessed this quality or, alternatively, she may have lost it in the course of time from disuse. In the Nile Valley the need for the honeybee to form a winter cluster never arises. When the Egyptian bee is crossed, the ability to form a cluster seems to dominate, but a colony of the pure *fasciata* cannot be overwintered with any certainty in Northern Europe.

When I set out for Egypt I feared it would be exceedingly difficult to find specimens of the pure *fasciata*, considering the wholesale importations that have been going on for nearly half a century. I soon learned differently. Whenever I came to a group of primitive hives, I found what was to all intents and purposes the pure native bee. One cannot well make a mistake in its identification, for the external characteristics of the pure *fasciata* are totally different from those of imported races. For some reason which has not yet been determined, the *fasciata* queen does not normally mate with drones of foreign races in its native habitat. Indeed, doubts were expressed whether it would cross at all, for some experiments carried out in Egypt seemed to show a physical inability. However, this is clearly not so, for crosses were obtained in Europe long ago. We secured cross-matings in the exceptionally unfavourable summer of 1963. But the fact remains that the ancient indigenous bee of Egypt has managed to retain its purity in its native habitat in the midst of imported races.

This descendant of the Pharaonic bee, of which we can still see representations made on Egyptian monuments as early as 3500 BC, will inevitably be doomed to extinction with the progress of modern apiculture. Thanks to her great vitality and the number of primitive hives, she has up to now managed to maintain her purity. The *fasciata* has not measured up to the needs of modern beekeeping, but this does not imply that she is of no value. I feel that every effort should be made, before it is too late, to retain this race at one of the numerous oases – perhaps at Sivas – for use in the future by the specialist breeder. It would be a major tragedy if this bee were lost to posterity.

This report would not be complete without a brief reference to the major problems which beset beekeeping in Egypt. This country is seemingly free of bee diseases, but the Egyptian beekeeper is faced with other problems no less formidable.

I have already indicated the great importance of cotton in the economy of Egypt, and the value of this plant as a major source of nectar. Unfortunately, the highly poisonous insecticides used for pest control are causing beekeepers immense losses.

The use of insecticides is a modern development, but the hornet (Vespa orientalis) is a menace as ancient as the Pyramids, and one found in all the countries adjoining the Mediterranean. Hornets cause serious losses in Cyprus and Palestine, but seemingly in no way approaching those in Egypt. It may be difficult to visualise the havoc wrought by oriental hornets without having oneself witnessed it. The honeybee can put up a successful fight against wasps but not against the oriental hornet, which seems to devour bees for a pastime. The damage to colonies would be catastrophic were it not for the counter-measures taken everywhere. Modern apiaries are usually provided with a battery of traps, similar to our wasp traps, but much larger. The primitive hives are generally fitted with small guards over the entrances, like our queen-exluders, but made of bamboo. This prevents hornets from entering hives, but in no way safeguards the bees when they venture outside. A person is therefore stationed all day amongst the hives, whose task it is to kill the hornets as they make a dash for the bees at the entrances. At one primitive apiary a rather different method was adopted: the cylindrical hives were left open in front so that the bees formed a solid cluster there covering the combs completely. Any hornet venturing near was overwhelmed by the bees, but individual bees could depart and alight directly on the cluster without risk of getting killed. It seems that hornets cannot catch bees in flight.

IN THE LIBYAN DESERT

I have already referred to the unique possibility oases can offer to modern apiculture for a progressive breeding scheme. The authorities in Egypt have been aware of this, and tentatively established a series of breeding stations at Kharga and Dakhla – the two large oases in the southern Libyan Desert. Though they are spoken of as two single oases, they are from our point of view two groups of a series of small oases. Kharga lies 145 miles by road south-est of Asyut; Dakhla is a further 125 miles to the west of Kharga.

A journey into the desert is no simple affair. Water, petrol and everything else has to be carried – including a competent mechanic in case of a breakdown. We set out with four vehicles and a party of 14 people. Amongst the party was Professor Baker of the USA, an authority on mites. He took advantage of this opportunity to include the two oases in his particular field of research.

The outward trek proved uneventful, though full of interest. Kharga was reached at night with the moon high in the sky lighting up the desert scenery. Kharga was already in ancient times a place of renown, as the Assyrian and Roman temple ruins bear witness. At present the oasis comprises six villages and a population of about 14,000. But an ambitious land reclamation scheme is under way here, which will soon change the existing state of things beyond recognition. The project depends on the existence and exploitation of the huge subterranean water supply below the oasis, estimated at 470 million cubic metres. This region is virtually devoid of rain; usually many years pass without any precipitation.

We spent one day at Kharga, exploring its suitability for breeding purposes. The Ministry had already established a small apiary for initial tests. Next day we set out for Dakhla, our main objective. The desert between the two oases is mountainous, rocky, very wild, and bare of all vegetation. However, at one spot by the roadside, a few dozen square yards in area, water seeped to the surface and a wild gourd grew in profusion. We were now approaching an oasis whose beauty of scenery and richness of vegetation and flora surpassed anything I had seen in Egypt, Algeria or Morocco. Dakhla, due to its remoteness and difficulty of approach, was until recent years cut off from the outside world. Our first call was at Tenida, where in a palm grove close to the village the Ministry had one of its bee-breeding stations. Here a great prosperity was at once visible. At the next station, about 15 miles further south, the richness and abundance were even greater. This station had about 40 colonies and, though all had two or three Langstroth supers, the hives were full up with new comb and honey. This apiary was barely a mile from the great sand dunes and the limitless ocean of sand. It was very strange to see these riches of newly-gathered honey, with the hives full of bees from top to bottom, in the month of December. I could also not help reflecting that here an Alpine bee had been brought into an environment as different as could be imagined from that of its native habitat. A more telling instance of the wonderful adaptability of the Carniolan bee – and of the honeybee in general – could hardly be found.

The same evening, though it was getting almost too dark, we inspected a third station, also with about 40 colonies; here we found the same prosperity and abundance. Our leader, Dr. Mohamed Mahmood, had, however, kept his gretest surprise until next day, when he took is to a small oasis called Rashda.

This one was undoubtedly the most idyllic and romantic of all, and the super-abundance and richness of vegetation and flora was truly astonishing. The Ministry's bee-breeding station was tucked away in an enclosure amongst date palms and orange trees, the latter laden with fruit.

I must remind the reader that the sole object in establishing these breeding stations has been to ensure pure mating – hence the large number of colonies. However, it is now realised that to obtain not only pure mating, but an improvement of strain as well, only the best colonies of a particular line must be retained at each station to furnish the drones. By limiting the efforts to pure mating, the primary economic benefits of breeding are completely missed.

THE SUPPLEMENTARY JOURNEYS

The main objective of the initial journeys was to ascertain the various individual races of the honeybee, their respective range of distribution and special characteristics.

During the intervening years each race and its varieties and strains were tested in the climatic conditions of South Devon, to ascertain their individual characteristics and breeding potentialities. In the course of the initial search I was of necessity compelled to base my assessments mainly on external characteristics and the behaviour of individual colonies. The likelihood that a particular colony would not necessarily embody the most valuable genetic dispositions of a race had constantly to be taken into account. The comparative tests alone, carried out in identical conditions against a series of races, strains and crosses, could provide the concrete facts and figures a successful breeding scheme demands.

The knowledge thus gained indicated the need of a series of supplementary journeys, confined to the areas where the most promising breeding stock had so far been found. There was, furthermore, every indication that in the areas in question even more suitable stock would be found than in the initial exploratory search. I had certain parts of Asia Minor, Northern Greece and the Sahara primarily in mind. The results secured on these supplementary journeys fully substantiated the foregoing assumptions and expectations and more than justified the additional outlay and effort entailed.

CONCLUSION

With the completion of these last journeys I have finished the task I set out to do. But our knowledge of the honeybee races is still far from complete. We know next to nothing of the indigenous honeybees of Iran and Afghanistan; nor have we much precise information concerning the economic characteristics of the races found in Africa south of the Sahara. Until these gaps in our knowledge have been adequately filled, any conjectures as to the origin of our present-day races will, I believe, lack a secure basis. The three Indian species, with the possible exception of *Apis indica*, have little bearing on our investigations. No doubt, with the growing appreciation of the more fundamental aspects of bee culture, and the research in progress connected with breeding and the genetics of the honeybee, someone with the necessary facilities and experience will take up the work where I have left off. In the course of my own search I covered about 82,000 miles by road, 7,792 by sea and 7,460 by air.

PART TWO

RESULTS OF THE EVALUATIONS

AN EVALUATION OF THE RACES AND THE CROSSES

The foregoing reports are an account of but the first stage of a very far-reaching venture. The second stage comprises testing and evaluating each of the races and the crosses resulting from them. Assessments of this kind demand of necessity that the experiments extend over a period of some years and that they should be carried out under certain definite conditions if the results obtained from them are to be reliable.

An evaluation can be made from three points of view, namely, to ascertain the possibilities of a race:

(1) when pure;
(2) when crossed;
(3) for the breeding of new combinations.

The improvements in the case of livestock and plants has been achieved either by way of line-breeding, viz. the intensification of the good qualities present in a particular race to the detriment of the undesirable characteristics; or by cross-breeding and the utilisation of hybrid stock; or lastly, by the more advanced method of combination breeding, the synthesisation of a series of good qualities, derived from a number of races, into fixed new combinations.

The real value of pure stock, whether formed by Nature or the influence of man, is only brought to light in cross-breeding. But in the case of the honeybee the best economic returns are not necessarily secure in a first-cross, as normally happens in livestock and plants, but more often in a second or back-cross. In the third method of breeding, involving the synthesisation of the good qualities possessed by different races, we have the means to form from races of little or no practical value fixed new combinations of outstanding economic importance. This last method of breeding enables us to utilise the immense fund of valuable qualities which Nature has placed at our disposal in the individual geographical races of the honeybee.

As you will see, we keep all these viewpoints in mind in our evaluations, for without doubt all three breeding methods will play an important role in the future progress of beekeeping. They are indeed complementary.

Essential requirements

As already mentioned for an assessment of the races of bees, well determined and reliable standards of comparison are necessary and this demands certain essential pre-requisities:

1. A large number of colonies.
2. A series of out-apiaries with varying honey-flow conditions.
3. The setting out of the hives in such a way as to prevent drifting, which might lead to misleading results being obtained.
4. Above all what is needed is a hive of a size to permit the maximum development of a colony, or more precisely, one which is completely adequate to meet the maximum fecundity of a race. The British Standards frame measuring $14 \times 8\frac{1}{2}$ in. and brood chamber holding ten combs of this size, as used by us at one time, is clearly not large enough for the majority of races. Performances obtained under such conditions are devoid of any real sound basis and tell us nothing.

The honey-flow conditions in which the evaluations are made should not be too favourable, for although consistently good crops can show what a race or cross can do, yet they do at the same time cover up hereditary weaknesses and disadvantages. We are here mainly concerned with the problem of disease and the hereditary dispositions of susceptibility and resistance. As our knowledge of this fact in bee-breeding increases, the greater emphasis is laid on it, and rightly so. We have to deal with the problems of disease and there is no sense in turning a blind eye to realities. The incidence of disease is greatly influenced by environment. From this point of view, as well as from other considerations, an environment with wide fluctuations, from extremely good to seasons of total failure, provide a safer basis for a positive evaluation than would be otherwise the case.

Where a number of races are being tested simulataneously – which is the only way comparative results are possible – they will not all show a like average in the amount of honey gathered. Those races which produce exceptional crops in a really good season may over a period of years fall far short of the general average results. Averages can be calculated in figures, but our estimate of the value of a race goes further and takes into account factors which are not reflected in the crops produced. In fact, they have nothing to do with honey production and their differences cannot be determined quantitatively. An example among many others is good temper, which cannot be measured in figures. In fact, it is a great problem in the evaluation of races of bees to determine the degree of excellence, but it is possible to lay down norms for judging qualities such as good and bad temper, although these norms are not mathematical. In all cases these norms are determined by comparison with a well-known race or strain, and this is a further essential requirement for a reliable evaluation. Our Buckfast strain ably fulfills this requirement for us.

I must here once more emphasise that from a commercial bee-keeping point of view, the touchstone of this assessment is the maximum average of honey produced per colony over a period of years with the minimum expenditure of time and money. This, however, does not always coincide with the breeding worth of any given race. There are races of bees which in favourable conditions produce amazing results, but in less propitious times will be a complete failure.

On the other hand, there are those which produce excellent averages over the years but at such an expenditure of time and money that they are not an economic proposition.

Our climate and honey-flow conditions

Our climatic and honey-flow conditions must be kept in mind for a proper understanding of the evaluation we have made of the different races. Although the essential characteristics of a race do not vary to any notable extent when they are placed in different surroundings, their advantages and drawbacks do show up more markedly.

In South-West England, where our work is being carried on, we do not have as a rule hard winters nor the long settled summers of the Continent. Our annual rainfall averages some 65 in. as compared to the average of 25 in. for the whole of Southern England. This, in addition, to the prevailing high humidity both in winter and summer makes our district less favourable for bees than other parts of the country. Total failures of the honey crop are by no means rare, while long periods of wet weather are a common feature of almost every summer.

Our main honey-flow is from white clover which yields in good weather from about the middle of June until the end of July. The heather, Calluna vulgaris, provides us with a second crop from the middle of August until early September, but the bees have to be taken to the Moor for this. Minor sources of nectar are willow in the spring, hawthorn, sycamore, fruit and blackberry, though the only fruit blossom we have is the apple. In other parts of England sainfoin and red clover provide a nectar source for considerable crops of honey, but these are not found in Devon.

This sort of climate and conditions of honey-flow demand a bee that must be above all proof against the winter, the weather and dysentery; it must be able to maintain its spring development in spite of unfavourable weather; it must have a flair for economy but at the same time the ability to meet every possibility of a honey-flow with colonies at full strength; it must be reluctant to swarm, and especially resistant to disease and notably acarine. Experience has shown over and over again that a bee which is at all susceptible to nosema, paralysis, or acarine cannot survive in our district. The year in, year out all pervading dampness, together with the lack of warmth and sunshine, demands a healthy constitution. Yet from the point of view of breeding these adverse circumstances have one great advantage, any susceptibility to disease or any other weakness show up at once.

RESULTS OF OUR EVALUATIONS

APIS MELLIFERA LIGUSTICA

It is doubtful if modern beekeeping would have made such tremendous forward strides in the past 100 years had it not been for the Italian bee. This bee is far from perfect, but she is endowed with a whole series of valuable characteristics which has been the cause of her world-wide distribution and guaranteed her a privileged position not enjoyed by any other race. She has her drawbacks, some of them very notable, which deny her an absolutely universal first-class rating. Her outstanding defects and the root of all her other weaknesses are her lack of vitality and her inclination to turn too much of the honey she gathers into brood. These disadvantages appear at their worst in the very light coloured strains of this race. The darkish leather-coloured bee which has its native home in the Ligurian Alps is without doubt the best of the many varieties.

The pure Italian does very well in favourable honey-flow conditions as found in North America and other parts of the world, but in conditions such as ours she often proves a dismal failure. Here she often requires feeding when strains less given to brood rearing can easily fend for themselves. Her tendency to turn every drop of honey into brood with all the spendthrift's disregard for any possible time of need manifests itself as one of her most unconomical traits in our changeable climate. Of course, this disadvantage must be accepted when there is a question of a late flow, as happens in our case. Experience has shown that the performance of this race on the heather and where red clover is grown is outstanding.

The lack of stamina with which the Italian is afflicted shows itself especially in the spring build-up, or rather in a failure to build-up, in a falling off in the strength of the colony. With us, spring begins at the winter solstice and five months later passes almost unnoticed into summer. It does not come suddenly with a burst of blossom as on the Continent, but rather creeps in almost imperceptibly, interrupted by considerable spells of cold and especially changeable weather, conditions which impose a severe strain on the vitality of the bees and can exhaust them prematurely.

I have the impression that queens which were imported from Italy before the First World War gave proof of a greater vitality than those which have been imported since then. In the meantime, apparently, the defects of the race have been intensified at the expense of its good traits. Today so much emphasis is laid on a bright yellow colour, but all my experience with this race clearly indicates that the unattractive leather-coloured Italian is by far the better bee commercially. A further drawback of this bee is its inclination to drift, caused by a deficient sense of direction. In fact I know of no other race in which this defect is so markedly developed, a defect which is of no material consequence when colonies are set out in groups as is our practice, but when set out in rows, with the entrances all facing in one direction, drifting can cause serious complications. In the case of beehouses, as in common use in Central Europe, drifting will be the cause of endless trouble. Indeed, the Italian been cannot be kept in beehouses with any measure of success.

On the other hand, the Italian possesses a combination of characteristics of great economic value. The near universal favour this race enjoys is clear proof of this. If managed appropriately both pure bred and crosses answer the needs of professional and amateur beekeepers as does no other race. In cross-breeding the Italian combines well with most races, both maternally and paternally – which can be claimed for few other races. In the synthesisation of new combinations, which is the breeding of the future, the Italian bee will clearly play an essential role, precisely because of its all-round compatibility.

The following points, however, need to be borne in mind: the greater number of the Italian's good characteristics are found in other races, where one or another of them shows itself even more noticeably, yet the worth of the Italian from the economic point of view is its possession of a combination of such a great number of these characteristics. Among them we can name: industry, good temper, fecundity, reluctance to swarm, speed in building comb, white cappings, storing honey away from the brood nest, cleanliness, resistance to disease, the preference for flower honey over honey-dew, a trait which I have noticed in only a few races and a factor of great importance in those countries where the light colour of the honey determines the price. Finally, as I have already mentioned, the Italian has shown its ability to gather good crops of honey from the red clover.

In one characteristic this bee seems to surpass all the other races, that is in its resistance to acarine. This fact was recognised by the authorities in England some 50 years ago, and it was on it that the government based its efforts to renew the bee population after the First World War. But it must not be assumed that all Italian strains show this resistance to acarine in equal measure. As a matter of fact, many of the present-day bright yellow strains have lost this quality completely.

I must mention in this connection our experiences with the strains which have been developed in America from Italian stock imported many years ago. Most of them have to be described as very bright yellow or golden, and all of these yellow strains which have been tried out in our apiaries have proved highly susceptible to acarine. Our experiences with these strains stretches back to 1924, but the high susceptibility I already then observed is no less marked today. It is rather curious that both American 'Goldens' and a strain of golden bees we developed from a French-cross, manifested this susceptibility in a way no other cross or race does, with the exception of the old English bee.

I ought to point out here that the course of this disease differs at least in our district from that shown by it more commonly. Heavily infected colonies collapse suddenly without warning, often in a matter of days, usually towards the end of July after a period of bad weather, though it can happen even in the middle of a honey-flow. But it is an insidious disease and losses of colonies during the winter and spring are of course common.

The re-population and change-over in England to Italians after the epidemic of acarine brought about an unprecedented rise in the average of honey produced, in spite of the drawbacks of this race to which I have referred. But I cannot really see this race standing up to the test either in a beehouse, as in common use in Central Europe, or where the main honey crop is gathered very early in the season as is the case in some districts and countries.

APIS MELLIFERA CARNICA

Apart from a few brief intervals we have had Carniolan bees in our apiaries continually since the turn of the century. During this time we have tried out innumerable strains which we have obtained from the many different districts in which this widely distributed race is found, and especially from those parts of Yugoslavia and Austria where it is recognised as existing in its purest form. In fact we have spared no effort to get queens from the remotest parts of Serbia and Montenegro so as to test this race as exhaustively as possible. As a result of these extensive experiments I have come to attach very great value to the Carniolan and I look for great things from it.

In the last 30 years this race has spread very widely in Central Europe, where generally speaking it has been accorded a very favourable reception. Indeed, it has been regarded on many sides, and with some justification, as the 'best bee'. In England, however, the opposite is the case, and I believe it is now to be found only in our apiaries.

The most important characteristics of the Carniolan bee from the economic point of view are her good temper, her extraordinary calmness during manipulations, her industry and stamina, her resistance to brood diseases, her keen sense of orientation; she makes a minimum use of propolis and seals the honey white; she is extraordinarily thrifty and comes through winter on a minimum of stores. However, her early and rapid spring development, for which this bee is noted, as also her limited fecundity, are of great value only in certain conditions. She is also very hardy and is endowed with a long tongue reach.

Among her undesirable traits are her extreme swarming tendency and, at least in our beekeeping conditions, her susceptibility to nosema, paralysis and acarine; finally, she is a decidedly poor comb builder.

In the Carniolan we have again a bee which is gifted with a long chain of valuable characteristics linked up with a small number of undesirable ones, but as is so often the case, these few bad traits exercise a dominant – an altogether too predominant – influence on the economic value of this race.

Without any doubt the Carniolan is the ideal bee for districts where there is an early flow from fruit blossom, dandelions, etc. In districts of this kind her early and rapid spring development can prove a most valuable characteristic. On the other hand, it is not easy to bring about the necessary harmonious balance between her twin tendences for a rapid build-up and her excessive swarming for her to make the most of a main flow in July, as is the case with us. Where there is a late flow such as from the heather, she falls far short of all the other European races I have tried.

Experience has shown that the characteristics I have described as of conditional value are of very real value for a honey-flow in early spring or summer, but where this is not the case they are a decided drawback. These disadvantages appear in worse form where a large brood chamber is used, as with us at Buckfast with our 12-frame MD hive. From my experience I would say that the Carniolan needs a brood chamber approximating to ten combs of British Standard size as used by us until 1930.

Until recently the Carniolan had the reputation of being prolific, but this is not borne out by our findings. After all fecundity is something relative dependent on the standard one adopts for assessing it. As was pointed out in one of the reports at the turn of the century, Chesire and Cowan, the pioneers of modern beekeeping in England, judged the Carniolan by comparison with the old English native bee, and their judgment has been accepted without further proof. It may well be that the very average fecundity of the Carniolan is an advantage in districts where there is an early flow, but we have found it to be a drawback in our honey-flow conditions, in fact, a very definite disadvantage.

According to our experience a Carniolan colony will rarely if ever have more than seven combs of brood of Modified Dadant size at any time. Furthermore, this bee reacts with great impetuosity to spells of unfavourable weather and will in times of dearth reduce or stop breeding altogether, even in the presence of an abundance of stores. This stop-go tendency in breeding, combined with the limited fecundity, has a very adverse bearing on the effective strength and honey-gathering ability of a colony.

Modern beekeeping, at least in English-speaking countries, demands above all a bee that is reluctant to swarm. A bee which showed no tendency to swarm would indeed be a godsend, not only for the commercial beekeeper, but also for the majority of small beekeepers. In our experience the almost uncontrollable swarming urge of the Carniolan is her most uneconomical trait. The widely accepted view that all she needs to put a successful brake on her swarming urge is a great deal of room affording her unrestricted possibilities for the free run of her comb potentialities has never received the smallest confirmation in all our experience. There are differences in the swarming tendencies between the various strains and the breeding material which we obtained from those places where beekeeping is confined to primitive hives. These last strains show an even more marked tendency to swarm.

Again we have not been able to confirm the widely held view that the Carniolan is an active builder of comb, at least in comparison with what our own strain is capable of. In our apiaries each colony must be able to draw out every year at least three foundations in the brood nest and practically all of the annually required combs in the supers, so there is no lack of possibility of comb building here. Indeed, we regard the reluctance of the Carniolan to build comb and the imperfect and irregular results of her meagre efforts as one of most noticeable features of this race, one which is manifested without exception in all the different strains. It is undoubtedly a contributory cause of and also a clear sign of the highly developed swarming tendency.

With the majority of races swarm control is always possible with the aid of the simple well-known methods. Moreover, even when swarming fever has got a hold, the normal activity of the colony proceeds as usual, although to a lesser degree of intensity. In these circumstances there is always a good possibility that the swarm fever will abate and disappear without any need of further action on the part of the beekeeper, and with few exceptions this normally happens. But with the Carniolan at the onset of swarm fever all useful activity ceases, and unless the swarming urge is given free rein or measures are adopted which we do not regard as economical, there is little hope of a reasonable honey crop even in the best of seasons.

As I have indicated, in spite of her many outstanding good qualities, the Carniolan is not suited to our conditions. Her very limited fecundity and her excessive tendency to swarm have their effect as is to be expected on her performance in honey production. Her abnormal swarming tendency alone causes such a waste of time in regard to the attention each colony requires that she is not a profitable proposition in a modern commercial undertaking. Whether it would be possible in the course of time to breed a strain in which this tendency was reduced to acceptable proportions is a moot point. Another factor which tells against her being used in modern commercial beekeeping is her constant demand for feeding during spells of unfavourable weather to keep her breeding, which does not fit in with present-day ways of management.

One of the most desirable qualities of this bee is its extreme gentleness and complete lack of nervousness. In behaviour she represents the extreme opposite of disposition to that of the common European black bee. The same holds good in regard to the use of propolis; the real Carniolan will use wax in place of propolis. Unfortunately, in the current commercial strains this unique quality has been largely lost, as also the disposition to cap the honey white – which I regard as two of the characteristic qualities of this race.

Although in our conditions, from a strictly business point of view, the Carniolan cannot compare with many of the other races, yet I regard her as absolutely indispensable for purposes of cross-breeding. Indeed, she is the key which opens up for us the potential of other races, especially those of the eastern group. My experimental work shows clearly that this bee in many ways is something of a puzzle; from the breeding point of view there are concealed in her as yet unknown possibilities which will emerge into the light only in cross-breeding. This, of course, is true in part of all races; cross-breeding brings out qualities unsuspected in pure-bred stock.

I must emphasise here: where there is a question of a general 'utility' cross with regard to this race, Carniolan drones should be used in every case. A reciprocal cross – Carniolan queens mated to drones of other races – produces very often a bad-tempered bee and almost invariably a first-cross of little or no economical value. Heterosis intensifies the swarming tendency to an even greater degree than is normally the case, when the queen is a Carniolan, with the result that such a first-cross expends all its strength in this craze for swarming. In addition, we are given here a classic example of the failure of heterosis to influence the already determined fecundity of the race, which is a further economic disadvantage of this sort of cross. Yet in the next and subsequent generations there is a marked decline in the swarming tendency, which allows for a full development of those characteristics which have a direct bearing on honey production, while at the same time a greater fecundity shows itself often to a far higher degree than manifested in the original parent stock.

It would be premature to assess the value of this race in the breeding and synthesisation of new combinations. However, there is little doubt that it will play a predominant role in this field, possibly a more predominant one than that played by the Italian.

Sub-varieties of the Carniolan

We know that the extent of the spread of the Carniolan comprises more or less the whole of South-East Europe. It is not surprising then that in this very large area, with its variations of climate and environment, there should be a number of sub-varieties of this race. For all we know, the Italian may well be a yellow form of the Carniolan. But of the sub-varieties of this race two come foremost to mind: the Banat bee and those which are found in the Carpathians.

We have tried out the Banat bee. Externally she is hardly distinguishable from the typical Carniolan, but she does differ from it in certain characteristics. For example, she propolises more than the ordinary Carniolan, and she builds a large number of queen cells at the onset of the swarming fever, which the Carniolan does not do. She is on the whole perhaps not so prone to swarm, but apart from this she possesses no characteristics of any economic value which is not present in the Carniolan in a more developed form.

As far as the bees of the Carpathians are concerned, I am not yet in a position to give any definitive report.

APIS MELLIFERA CECROPIA

The native bee of Greece undoubtedly belongs to the same family circle as the Carniolan, yet she is rightly regarded as a special race, since she differs from the Carniolan in a number of essential characteristics. But even within the frontiers of Greece itself different varieties of the race are found. In my view the varieties found east of the Pindus mountains from Attica to the northern frontier of the country are economically the most valuable. Hence my findings are restricted to these. It seems to me in view of the conditions prevailing in the northern districts of Greece, that it is very questionable whether the description 'Macedonian' for the bees there is correct.

Until I gave my lecture at the International Congress in Vienna, practically no attention had been given to this race. The Greek bee has, indeed, none of those external features which would attract attention; she has neither the light colour nor the uniformity of colour on which so much value is placed. Yet in spite of her unattractive external appearance the Greek bee, in my view, has hardly any equal from the economic and breeding point of view. I realised this at the time of our first importation of these bees in 1952, and since then this view has been greatly strengthened.

There are no marked differences externally between the Greek bee and the Carniolan, apart from the tendency for an occasional leather-coloured band to show up more frequently. Nor is there any material differences in gentleness between the two races but, in fecundity, the Greek surpasses the Carniolan and in her reluctance to swarm no other race can seemingly equal her. As for actual colony strength few can compete with the Greek, especially when queens of this race are crossed with Italian or Carniolan drones. The strength attained by such colonies is quite phenomenal. However, longevity is in part responsible for the great colony strength. Exceptional thrift is another quality of great economic value which the Greek shares with the Carniolan.

Although Greek queens are prolific, they are not more so than the common Italian. Moreover, this race will not breed out of season to excess. Indeed, breeding is severely restricted at the conclusion of the main honey-flow and in some strains this tendency manifests itself more than we would wish. Due to this highly developed thrift Greek colonies will as a rule demand far less feeding than Italians in identical circumstances.

Great fertility and colony strength when not allied with reluctance to swarm, as so often happens, would of course be of no real advantage, at least under the conditions in which we work. A proclivity to swarming renders useless any gain from an above average fecundity. Get the two together and we have the basis for a highly productive and profitable beekeeping. In the fact that these two most important qualities are linked in the Greek bee, I see the real value of this race both commercially and from the breeding point of view.

As regards her less agreeable qualities, the Greek closely resembles the Anatolian and Caucasian, especially in her excessive use of propolis and construction of brace comb, and her watery, flat cappings. But these failings are far less prominent in the Greek, indeed, there are strains in which they hardly appear at all. In certain crosses these defects vanish altogether and rather astonishingly a type of capping of identical pattern and perfection as made by the old English bee will now and again manifest itself. Regarding disease, the Greek seems much less susceptible to nosema than the Carniolan, probably because of the great colony strength with which she comes through winter and a less precipitate brood-rearing in the early days of spring. I have never observed any sign of acarine among them, though paralysis will show itself in certain strains. This last mentioned susceptibility manifests itself most when there has been inbreeding to any extent. Indeed, as far as my experience with them goes, the Greek bees are more sensitive to inbreeding than most races.

I have already mentioned the great strength with which Greek colonies come through winter. Although the spring development is not so rapid as with the Carniolans, yet it is more than ample to take advantage to the full of an early flow. As a matter of fact, due to this unusual colony strength and reliable spring build-up this race is particularly suitable for the pollination of fruit and where there is the possibility of an early honey crop.

With the exception of the white cappings and especially the non-use of propolis – two characteristics which, unfortunately, are largely lacking in the modern Carniolan strains – the Greek bee has the majority of the good qualities of the Carniolan, plus the additional qualities of fecundity and reluctance to swarm, traits which we miss so much in the last named. Furthermore, these additional qualities can be put to the best possible use in cross-breeding. In the case of the Greek bee reciprocal crosses prove equally advantageous. In fact, as our comparative tests have shown, a first-cross – either Greek queens × Buckfast drones or vice versa, as well as a back-cross to Greek or Buckfast drones – produces a bee which from the practical and technical points of view can hardly be surpassed in general utility and performance. This holds good not only where pollination or a crop from early sources is a primary consideration, but equally so where the main crop is derived from the clover or a late honey-flow such as the heather. My experience has been mainly restricted to these crosses, but I have little hesitation in saying that similar results could be obtained with other crosses, especially with drones of a proven Italian strain. The reluctance to swarm is dominant even in the first-cross with Carniolan drones. This dominance is most surprising in view of the fact that in the majority of crosses in all races heterosis is inclined to bring about the opposite results, namely, an extreme tendency to swarming. Good temper, fecundity, self-sufficiency and also to some extent thrift, exert an equal dominant influence in a cross of this kind.

The Greek bee has from every standpoint a great future in front of her. Unfortunately, in present circumstances it is not easy to obtain first-class breeding stock of this race. In her native land breeding and selection are still largely left to Nature.

APIS MELLIFERA ADAMI

On my first visit to Crete in 1952 I left this island with the impression that the honeybee found there was extremely aggressive and that it did not possess any characteristic meriting special attention. However, the biometric studies carried out by Prof. Ruttner on the samples of bees collected by me, indicated that the honeybee of Crete constituted a distinctive race and one endowed with a series of unusual external characteristics. He named his newly-discovered race *Apis mellifera adami*.

Subsequent to the publication of these findings, I deemed it desirable to give this hitherto unknown race a comprehensive test in our climatic conditions. The unusual aggressivity of this race proved itself if anything more pronounced in our climatic environments that when first observed in its native habitat. However, we were able to ascertain certain physiological characteristics which eluded the more cursory observations made in Crete. One of these is the disposition to raise an immense number of queen-cells in compact clusters on worker brood, with cappings closely resembling those of drone brood. But individual queen-cells are likewise produced in great profusion, resembling in size and shape those of the Egyptian bee.

The queens raised from the pure Cretan stock were duly crossed with our own drones. A first-cross is usually more aggressive, due to the heterosis, than the parental stock. To our surprise, the reverse was the case here in most instances. The first-cross proved remarkably gentle, exceptionally thrifty and outstandingly good honey-gatherers. A series of further tests confirmed the initial findings. We feel convinced this race will prove of great value in cross-breeding and wherever adequate control of the drones is assured. Random matings should in no case be accepted.

APIS MELLIFERA CAUCASICA

I have not yet had an opportunity of visiting the native habitat of this bee, but this in no way implies lack of interest on my part. On the contrary. It is now more than 30 years since I first imported Caucasian queens. These came from North America. In the meantime I have tried out strains from a number of sources of supply including queens direct from the native land of this race. I have never had much success with this bee. However, from the particulars given in a number of reliable reports there are presumably strains capable of excellent performance, although I sometimes wonder if the comparisons underlying the results of these reports have not been made on too narrow a basis. At all events there is no doubt that the name Caucasian covers a whole series of strains of very dissimilar economic worth. Moreover, some of the commercial varieties clearly do not represent the true Caucasian.

In externals, colour and grey over-hair and tomenta, as well as good temper and tongue-reach, this race is very similar to the Carniolan, and in the last two qualities she is noticeably superior. Against this the Caucasian goes to extremes in building brace-comb and use of propolis. In regard to these two most undesirable traits she surpasses every other race, although in the use of propolis there are some very close behind her. The drawbacks of these two traits are intensified by mutual interaction, and as in the Caucasian they are developed to an abnormal degree; this makes manipulation of them in a modern hive very difficult. Hence, in spite of their many good qualities, these two undesirable dispositions have set serious obstacles to a wider distribution of this race. In the Caucasian the excessive use of propolis is furthermore such a highly developed characteristic that it is transmitted in an almost unmitigated intensity from generation to generation in cross-breeding. Whereas the tendency to build brace-comb is fairly easy to breed out, that for propolising is in every case transmitted as a dominant factor and can only be eradicated at the cost of endless trouble.

The Caucasian is universally considered as the most gentle of races, but there are strains which are anything but good tempered.

Apart from this trait of extreme gentleness, this race is noted for its exceptional tongue-reach which in certain strains of this bee attains the highest average of all races. However, it must not be assumed that honey crops from red clover are directly proportionate to tongue-reach; that is, that bees with the greatest tongue-reach will infallibly produce the most honey from red clover. The majority of Italian and Carniolan strains give equally good crops from this source.

This race is too classic an example of a bee which will in most seasons provide its own winter stores and will normally store the newly-gathered honey within a minimum of comb space. The advantage of this latter trait is that at the end of the flow or when a flow suddenly ceases, there are not an undesirable number partly filled and unsealed combs left on hand. This tendency, which can be also observed in the Carnolian but not in such a marked form, makes for a higher quality of honey especially in a very damp climate such as ours.

As regards fecundity, we have so far discovered no essential difference between the pure Carniolan and Caucasian. The latter reacts to a pause in the honey-flow in exactly the same way as the Carniolan, that is, there is a sharp fall in brood rearing. In our climatic conditions the Caucasian has also shown a susceptibility to acarine and nosema and, in general, it is not a race as hardy as would be expected from a mountain bee.

Although this race possesses a number of valuable traits it is not well suited for cross-breeding purposes. None of the crosses we have been able to put to the test have been satisfactory. To sum up: the characteristics of the Caucasian which are desirable for our present purpose can be obtained in other crosses without the disadvantages of the undesirable traits of this race.

APIS MELLIFERA ANATOLICA

As I have pointed out in the reports of my journeys, Asia Minor is the home not of one race but, apparently, of a number of races of bees, and as one would expect, in the districts bordering the habitat of these races there is a whole host of strains of intermediary origin. To complicate matters still further, there are islands of one race in the middle of an area populated by another. In fact, it is often difficult to determine where the most typical examples of one of the indigenous races can be found.

The dark bee of the north, that is, the district east of Sinop, shut in between the Black Sea and the Pontus mountains, differs considerably in its behaviour and economic characteristics from the Caucasian. Similarly, the orange coloured bee of the former Armenia differs from that of Central Anatolia, which both in colour and other characteristics can be regarded as a fixed intermediary form between the two above-mentioned races. The bees of Cilicia, domiciled in the narrow strip of land between the Taurus and the Mediterranean, resemble the Syrian bee in external markings, but are very different from it in most other respects. The specific differences between the races, both in behaviour and in physical traits, often show themselves here as in other cases more clearly in a first-cross rather than in pure stock.

All the above-mentioned races with the exception of the Syrian, of which only intermediary forms are found within the frontiers of Southern Turkey, have certain characteristics in common, although these are very differently emphasised because of the different environment in which they have their habitat. They are all very thrifty, but as would be expected the Cilician the least of all. For good temper, the Pontus bee leads the way, while the Cilician is at the other end of the scale. But there are strains of either race which can be described as bad tempered or as extremely good tempered. They all have one thing in common: they are all sensitive to the cold, which shows itself in unusual bad temper whilst such conditions last. This tendency is found in all races, but never so marked as in the Anatolian varieties.

In comparison with other races, the pure Anatolian varieties are below average fecundity with the exception of the Cilician. None of the others come up to the standard set by the Carniolans. Contrary to these the Anatolians, when crossed, are prolific to an almost unbelievable extent, although at the same time all of them except the one from Central Anatolia show a marked tendency to swarm. Even in the case of pure stock, there are very noticeable differences among them. For example, the Armenian variety will build an enormous number of queen cells. 200 or 300 cells is not uncommon, and in spite of this huge number the young queens are as perfectly developed as they could be, without the slightest sign of any undernourishment or defect of any kind.

I have already referred to their sensitiveness to cold, but this shows itself exclusively to an increased transient stinging propensity and has no bearing on their wintering ability. In fact, as far as wintering is concerned, the Anatolians are superior to all other races known to me. The Armenian variety holds the first place easily. In the exceptionally cold winter of 1962/63, which was the coldest in the South-West of England since 1750, we wintered miniature colonies of pure Central Anatolian bees on six combs – $7\frac{1}{4} \times 5\frac{3}{4}$ in. – with complete success, a feat which seemed scarcely possible in the circumstances.

I mentioned previously the longevity of the queens, a characteristic which without doubt has a corresponding influence on the length of life of the bees and hence on the ability to winter well, and also in the attainment of an optimum effective colony strength. Taking into consideration the facts about their fecundity, their exceptional colony strength would be inexplicable without their vitality and power of endurance.

Another remarkable characteristic of this group of races is their highly developed sense of direction or orientation. Apart from an absence of drifting, this trait shows itself most strikingly in a low loss of queens when returning from their mating flights. Over the years we have calculated such losses in our own strain as about $22\frac{1}{2}$ per cent; in the Carniolan 10 per cent; but in the Anatolian and Middle Eastern races only 5 per cent.

In the performance or honey-gathering ability of a colony or race a whole series of characteristics is reflected. It is not just a question of zest for gathering the nectar. In the Anatolian bees we have a combination of the required factors as found in few other races. But among the varieties and strains of this group of races there is, nevertheless, a considerable variation in honey production, especially in the first-crosses, due partly to the difference of the tendency to swarm. We have had by far the best results from the variety which has its native habitat in Central Anatolia.

Of the factors which have a negative influence on production the chief one, apart from the swarming tendency, is susceptibility to disease. Curiously enough the Anatolian group of races is particularly prone to paralysis. We know that this disease is caused by a virus. But this virus cannot cause the disease unless there is a predisposition to it present in the bees, which experience has shown to be hereditary. In our environment paralysis makes its appearance at a certain period in spring, but once this time has passed no sign of the disease is seen for the rest of the year. In Turkey itself I never saw any sign of the disease, and there are strains in which paralysis never appears. I need not stress, of course, that paralysis affects all races of bees, but seems to affect the Anatolian group to a much greater degree than the others. Very fortunately this predisposition can be bred out relatively easily. The Armenian variety appears to be subject to acarine and nosema, but I have so far not come across any other special susceptibility or resistance to disease amongst the Anatolian varieties.

This group does, however, suffer from another troublesome disability, namely, a failure to ripen fully the nectar from the *Calluna vulgaris*, with the result that in some years the heather honey ferments in the combs a few days after it has been capped and sometimes even before it is sealed. But according to our experience this inability occurs in all races of bees, least of all in the Carniolan and Intermissa group. With one exception it shows up worst in the Anatolian group, although even here there are wide variations. Here again experience has shown that with careful selection and breeding this defect can be eliminated.

In addition to the susceptibility to paralysis, the defect concerning the ripening of heather honey, the susceptibility to cold and the tendency to swarm, mention must be made of two further undesirable dispositions: the inordinate construction of brace-comb and use of propolis. The last trait is not so pronounced in the Anatolian group as in the Caucasian. When they are crossed appropriately, the hybrids show these defects in a very mitigated form, and with proper selection completely disappear in a few generations.

One thing is clear: it is impossible to find a description which will adequately cover this group of races. For it is a group.

Granted that there is a close inter-relationship, the difference between the races of the group – and by differences I do not mean merely external markings but differences of behaviour and physiological traits – are as clear and of no less importance than those between the races long since recognised as distinct. In fact it has not yet been fully established how many really distinct races there are within the confines of Asia Minor.

The bees of Central Anatolia have so far given us the best results, and I have given full details of these results in the appropriate report of my journey. This race is also found in the districts adjoining Anatolia on the west and south-west, but the bees differ somewhat corresponding to the changes of environment. Comparisons we have made show that these other races are superior to the Central Anatolian in certain features, but taken as a whole, the Central Anatolian is economically the best and most productive bee. Further experience has also confirmed that my report summed up accurately the essential traits of the Central Anatolian bee, and that there is nothing worthy of note to add or or subtract from it. There is one exception to this affirmation, but it does not touch on the evaluation of the race.

My comparative tests have been limited to crosses with Buckfast bees in the main. At one time I was of the opinion that similar results would be obtained by crossing Antolian queens with Carniolan drones. It appears, however, that this is not so. As I have already mentioned in another connection (and wider experience has confirmed it), a cross with the Carniolan or with the Caucasian in many cases does not produce a good tempered bee, at least in the first-cross. Similarly, a Buckfast/Anatolian first-cross gives us a bad tempered bee, but at the same time one of outstanding productive ability. However, the reciprocal cross – Anatolian queens × Buckfast drones – is even more productive. In these two crosses heterosis does not accentuate the swarming urge, and the Anatolian/Buckfast cross is much more prolific.

Again I must emphasise that with the pure Central Anatolian bee as, indeed, with this whole group of races, it would be futile to expect maximum performance. The really economical valuable characteristics manifest themselves to the full only when suitably crossed. Where there is no control over the drones, random matings should be strictly avoided as an unsuitable pairing can produce an extremely bad tempered progeny.

APIS MELLIFERA FASCIATA

In the Egyptian bee we have a race of exceptional uniformity and distinctiveness. From the earliest times it has been restricted to the narrow limits of the Nile valley and completely cut off from the outside world, circumstances which have given it the seclusion necessary to develop a quite remarkable uniformity. From the economic point of view this race is of no importance, but invaluable for cross-breeding purposes.

The essential external features and characteristics of the Egyptian bee have been noted in the report of my visit to Egypt. However, I would like to refrain for the time being from particulars of its reaction to our climate and beekeeping conditions and the results obtained in cross-breeding, as there have been a number of special problems in carrying out experiments with this race, and I should like more time before offering any reliable data.

The Egyptian bee is, undoubtedly, one of the primary races from which came the orange coloured varieties of the Near East; that is, the Syrian, Cyprian, Cilician and, possibly, also the Armenian. In any event the influence of the Syrian is clearly noticeable in the far eastern regions of Asia Minor.

Supplementary findings

The Egyptian honeybee was until a few years ago known as *Apis mellifera fasciata*. But due to the discovery that this name had been bestowed by Linnaeus on another insect, the Egyptian bee was renamed *Apis mellifera lamarkii*. However, I am retaining the former designation, for this race is best known as the Fasciata. There is no likelihood of this causing a confusion; should this happen a change would then be fully justified.

Our cross-breeding experiments, incorporating this particular race, have revealed many interesting results. In keeping with the disposition of the pure Fasciata the F-1 proved usually bad tempered. The subsequent generation were, on the other hand, exceptionally prolific, good tempered and remarkably quite even when manipulated. Indeed, so much so that queens of Egyptian lineage will often continue laying whilst under observation – an occurrence one can rarely observe in the case of any other race of cross-bred stock. A Fasciata cross will, however, manifest a deficient ability in coping with harsh inclement climatic conditions. This is not surprising for in its native habitat it is not subjected to a need of this kind and therefore lost its ability to form a winter-cluster. But as we have found, these deficiencies can be progressively eliminated in cross-bred stock. The pure Fasciata possesses one particularly unique characteristic, namely, it does not collect any propolis. Apart from the Indian species it is the only race of honeybees endowed with this disposition. It is one we value highly, but it is also a rather elusive quality and recessive to the many genetic factors that determine propolisation. On the other hand, we have good reason to believe, that we will in time manage to isolate this highly desirable trait of the Egyptian bee in a new combination.

APIS MELLIFERA SYRIACA

The Syrian bee and the Cyprian are often regarded as one race. True they have many characteristics in common, both good and bad, but there are very marked differences between them, and anyone with wide experience of both races can easily distinguish one from the other. It is more than likely that they are closely related. In fact we have good grounds for surmising that the Cyprian and the Cilician varieties are descended from the Syrian, and the Syrian in its turn is an intermediary form between these two races and the Egyptian.

The Syrian is one of most interesting and attractive bees. In size, colour, in its white or almost white hair, in its sensitivity to cold, in its sparing use of propolis and in other characteristics, this race is closest to the Egyptian, and this shows itself most strikingly, apart from size and colour, in its sensitiveness to the cold. The Syrian becomes benumbed at temperatures in which the Cyprian is still very active. As is to be expected this sensitivty has an adverse influence on its industry and performance in northern latitudes.

The extreme bad temper of the Syrian is probably due in large measure to this sensitiveness to low temperatures. It definitely makes her irritable. Not that she is of the unprovoked, attacking type. Indeed, she will not attack except when roused by a disturbance of her home, but once her stinging propensity is inflamed, this bee knows no limits. Very strangely this fierceness of temper in the Syrian can be easily eliminated in cross-breeding.

Since the Syrian bee has no claim to be of economic value in her native land, it has still less in any other land. Even for purposes of cross-breeding, I see little future for it. The good characteristics the race possesses are found in much more useful form and to a greater intensity in the Egyptian, the Cyprian or the Cilician.

APIS MELLIFERA CYPRIA

My experiences with the Cyprian bee date back to 1920, although even ten years earlier than that there was a Cyprian cross in our apiary which had been raised by Samuel Simmins, who co-operated with F. R. Cheshire and the American F. Benton in importing the first bees of this race to England. But it was not until 1920 that the first queens reached us direct from Cyprus. They came from the neighbourhood of Nicosia. In the summer of 1921 we tried out more than a 100 colonies headed by queens of this race crossed with Italian drones. It happened to be a very good honey year, and the good and bad traits of this bee were soon apparent. From then on we imported queens from different parts of the island, and in May 1952 I had an opportunity of seeing this bee in her native habitat.

In Northern Europe, as elsewhere, the pure Cyprian has little claim to be of any exceptional economic value in spite of its many good qualities. The amazing fecundity of the Cyprian is well known, but it lacks the corresponding thrift and, consequently, more honey is used for brood rearing than the beekeeper desires. Its industry is unequalled, but there again it is not suited to an early flow or one late in the season, since the honey gathered early in the season is turned into brood, and as there is a sharp fall in colony strength after the main flow, the results on the heather are far from satisfactory. Pure Cyprians are not greatly given to swarming, but no matter with what other race they are crossed, the first-cross is very prone to swarm. However, should a Cyprian cross be faced with a good flow, the swarming will abate overnight and a good crop of honey results.

It is often assumed that a race coming from the sub-tropics would necessarily lack hardiness. This holds good in some instances, foremost in the Egyptian bee. But in the Egyptian the inability to winter in our northern regions is caused by a dual deficiency, namely, a want of stamina and an absence of a disposition to form a winter cluster. Though related to the Egyptian the Cyprian, either pure or when crossed, will on the other hand surpass in hardiness and wintering ability every other variety of the honeybee. In the severest of winters and most adverse spring conditions, as experienced over a period of 50 years, we have never had a colony of this race or cross that failed us in regard to wintering or one that failed in the spring build-up. This is clearly due to the amazing vitality with which the bee is endowed. However, an extreme vitality of this kind is liable to have drawbacks in other directions.

Nothing has brought this race more into disfavour than its excitable temperament. And rightly so, as the majority of the strains react, especially in cold or unfavourable weather, to any interference with a truly astounding energy. Moreover, its stinging propensity is not limited to dealing with any disturbance in the neighbourhood of the hive, but extends to a merciless pursuit of the intruder for a considerable distance. This most undesirable trait she shares with the Syrian. It must, however, be pointed out that this display of temper appears only when dealing with interference or disturbance of the hive. In Cyprus I often saw large collections of primitive hives in gardens and courtyards, surrounded by houses, where people were continually passing without anyone being molested by the bees. This is a clear indication that the bee is not given to spontaneous unprovoked attack which is, for instance, such a characteristic trait of the common black European bee.

It could be concluded from its ability to winter well and build up quickly in the spring irrespective of the most adverse conditions that this race is particularly resistant to disease, at least to those diseases which affect the adult bees. And this is indeed the case; I believe we have here a bee that is unsurpassed in this matter. At the same time I have never yet seen any such defects in the brood as occur here and there in other races. On the other hand, one very undesirable trait shows itself, but one which has nothing to do with disease, is in fact another indication of the extreme vitality of this race, namely, the appearance of laying workers within a short time of the loss of a queen. This happens whether the bees are pure or crossed and the Syrian bees are equally susceptible to this defect.

There are some other characteristics of the Cyprian which ought to be mentioned. When dealing with the Caucasian and the Anatolian group of races I deplored their extreme tendency to build brace-comb. All the races tend to make use of brace-comb in varying degrees. The Cyprian, however, seems to lack this disposition completely. The absence of this trait has no bearing on the honey crop but is, nevertheless, very advantageous. The presence of brace-comb can undo to a very great extent the essential advantages of a modern hive, or at least make the removal and handling of combs a very difficult and unpleasant task.

The Cyprian possesses to an unsurpassed degree a sense of direction and orientation; we have abdundant evidence of this in the low loss of queens when returning from their mating flights. On one occasion, from a batch of 110 Cyprian virgin queens only one was lost, and that at a time of the year when losses are generally above the normal average. A highly developed sense of smell is doubtless a necessary pre-requisite for an above-average sense of direction; it is probably inseparably bound up with it. They are, indeed, complementary. The traditional arrangement of the primitive tubular hives in Cyprus is to place them in four or five layers, one on top of the other, in stacks of great length with scarcely any kind of distinguishing mark. An arrangement of this kind necessitates a faultless sense of orientation and recognition. But a very acute sense of recognition has its drawback. As is well known, such bees are very difficult to unite. Our tests have shown that these traits are not the sole prerogative of the Cyprian bee but are equally a mark of the whole Fasciata group and that of at least of some of the Anatolian varieties.

There is no doubt that many valuable traits of the Cyprian bee appear at their best only in cross-breeding. The hundreds, nay thousands, of years of inbreeding, within the limits of a comparatively few colonies have covered up the full potentialities of this race. The complete isolation of the island, the inbreeding over thousands of years, the hard living conditions, the scanty resources, the merciless natural selection have conspired together to bequeath to us a bee of inestimable value for cross-breeding. But the Cyprian is not the type of bee the commercial beekeeper or amateur can make profitable use of.

APIS MELLIFERA INTERMISSA

The Tellian bee, the native bee of Tunisia, Algeria and Morocco, is another of the primary races. This coal-black bee has a bad reputation, and in my opinion she is not to be recommended to either the amateur or the commercial beekeeper. However, although her direct economic value is very small, I am convinced the race has a valuable part to play in cross-breeding.

The genetic make-up of this bee offers great possibilities, for better and for worse. She is well known for her extreme bad temper and nervous disposition, though there are strains which can be handled with impunity. But in my view her worst characteristics are her unbounded swarming tendency and uninhibited disposition to rear brood in and out of season, and complete lack of thrift. In South-West England these meaningless tendencies manifest themselves as late as the end of September at a time when with other races most colonies have no brood at all. Indeed, I have now and then been forced to take away the queen from colonies of this race while we were feeding them for winter so as to prevent all the stores being turned into brood. Extravagant dispositions of this kind tend to render any other economic traits useless; all the good traits are there that make for first-rate performance, but their value is all frittered away in wanton swarming and breeding. There is in fact no middle course with this bee; she goes to extremes in everything, a carefree spendthrift, a child of the wild, but gifted with the primitive exuberance of energy and vitality.

This race and all its sub-varieties labour furthermore under a very marked hereditary weakness. In the case of the Carniolan I drew attention to its remarkable freedom of hereditary defects and diseases of the brood. Here we have the opposite extreme, a quite extraordinary susceptibility to brood diseases and defects of the brood. This susceptibility shows itself very strikingly in the native land of this race, in fact brood diseases constitutes the chief danger to any profitable beekeeping in North Africa. We know that in general these and other defects are the results of a lack of vitality, and inbreeding is often a pre-disposing cause. But this is defintely not the case here. It is due primarily to a special inherited defect and susceptibility, as I have been able to verify again and again. As regards diseases affecting adult bees, the Tellian has stood up well to nosema; I have never observed a trace of paralysis, but she is very prone to acarine.

When dealing with the Anatolian group of races, I mentioned the peculiar inability of a number of races to ripen properly the nectar of *Calluna vulgaris*, at least in certain climatic conditions. I have up to now never seen a sign of this defect in the Tellian or its sub-varieties.

One other noticeable characteristic of the Tellian is its highly developed urge for gathering pollen. She is really a glutton for pollen as no other race is. This trait is one running right through all the sub-varieties which owe their origin to the Tellian. On the contrary, as is well known, the yellow races in particular do not gather correspondinly large stocks of pollen. Indeed, in their case we rarely if ever see combs solidly full of pollen.

The Tellian is not suitable for first- and second-crosses that are to be used for the production of honey. The extreme vitality and swarming disposition are intensified even more by the heterosis, with the result that their economic value is completely lost. This, however, does not mean that the race is useless.

On the contrary, the task of modern bee breeding is to harness the primitive vitality hidden in this bee and incorporate it together with her many outstandingly good characteristics in such combinations as will best serve the requirements of the practical beekeeper.

SUB-VARIETIES OF THE INTERMISSA

As I have already indicated, there can be no doubt we have in the North African bee a primary race of which the numerous sub-varieties have spread via the Iberian Peninsula, Central Europe and Northern Asia, possibly to the far distant shores of the Pacific Ocean. All the evidence we have supports this view and anyone who is acquainted with the primary race as well as the West and Northern European varieties, can easily trace its characteristics as they show up in the different varieties. As is to be expected all these characteristics, both economic and uneconomic, are developed to the highest degree in the parent stock. On the other hand, we find in the West and Northern European varieties a progressive graduation. I shall limit my comments to those varieties which have in the course of the years been tested and about which we have comparative records.

I might well discuss the peculiarities of these sub-varieties under the one heading, for in every case we are mainly concerned with progressive modifications of the primary characteristics. But from the point of view of the practical breeder it will be of assistance to indicate the particular qualities and drawbacks of each variety.

Apis mellifera major nova

The particular honeybee found in the Rif mountains of Northern Morocco is doubtless a local variety of the Intermissa. Its exceptional external or morphological characteristics differ from those of the race it originated only in regard to size and in no way qualitatively. The region where it is found is limited to a small area within the actual habitat of the Intermissa. Apart from a maximum known size of body, tongue reach and length of wing, the Rif bee is according to our findings identical to the Intermissa in its physiological traits and behaviour – with possibly one exception, namely, in the consumption of stores in winter. According to our findings, the pure Rif bee and crosses manifest an almost unbelievable extravagance in this direction. In identical circumstances, viz. locality, time and climatic conditions, the consumption of the Rif colonies averaged 14.4 kg.; that of an Anatolian cross 6.75 kg. The average of all other strain and crosses was no more than 9.45 kg. The Rif F-1 manifested in a marked degree the particular aggressiveness and undesirable traits of the typical Intermissa. However, notwithstanding its numerous undesirable characteristics, the Rif bee is at the same time endowed with certain traits in an intensity not found in any other race of the honeybee. This variety may therefore in conjunction with other characteristics serve a valuable purpose in breeding.

The Sicilian bee

The Black bee of Sicily, also known as the Sicula, is the nearest relative and the one most like the Tellian. In fact there is not a very notable difference between the two races. The seclusion and isolation of Sicily, together with the similarity of its climate to that of North Africa, precludes any likelihood of marked differences in the characteristics and it possesses no special points of any value.

The Iberian bee

As far as external markings are concerned, the Iberian bees differ very little from the Tellian, but in other respects there are very noticeable differences. In place of the spendthrift waste of stores caused by the unbridled rearing of brood, the unbounded swarming tendency and the corresponding useless waste of energy, the

Iberian bee has developed a very definite sense of good housekeeping. The great vitality and fecundity of the Tellian is still there, but there is a marked falling off in the tendency to rear brood out of season as called for by the needs of the environment and a regard for future requirements. As a result of these changes we have a variety of the Tellian which is capable of really good performance. However, the excessive use of propolis continues in unabated measure and likewise the stinging propensity, so that while this variety manifests certain modifications, the distinguishing features of the parent stock are still present quite unmistakably.

One of the most unfortunate and adverse features of this whole group of races in my view is their extreme nervousness and antagonistic attitude to the queen. In dealing with the Iberian strains and their crosses this trait was strongly impressed on me. Since the disappearance of the old English bee, nearly 50 years ago now, the balling of a queen is practically unknown, but with colonies of Tellian stock it is an almost daily occurrence unless one uses extreme care in handling them. Where at one time there was no difficulty in getting queens accepted, we are now constantly faced with exceptions to the general rule. It is not merely a question of acceptance of queens, but this group of races is quite unique in its whole attitude towards queen introduction. For example, it is not unusual to find in the months of June and July that a queen has been accepted unhurt, but a period of three to four weeks elapses before there is any sign that she has been accepted. Only then does she begin to lay. What is worse, there are all too often cases when colonies of this race group refuse to accept any queen at all which, of course, spells ruin for them ultimately.

A further bad feature of this race, which in the sub-varieties is most pronounced in the French bee and to a lesser extent in the others, is the tendency to build an excessive amount of drone comb – a trait which comes to the fore in the most accentuated form in a first-cross; that is, French queens crossed with other races. A reciprocal cross does not usually manifest this tendency. In the former type of first-cross a high percentage of foundations are inevitably ruined. This urge to build drone comb, although present in almost all first-crosses, is most accentuated in the Iberian and French varieties.

The whole Tellian group is undoubtedly highly susceptible to acarine. As far as I have been able to ascertain, of all the sub-varieties the Iberian bee shows this defect most, not only in our climatic conditions, but in its native land as well.

The Iberian bee has made great strides away from the present race, and she is a variety capable of giving good results. If anyone desires a pitch-black bee as his ideal, he will find his desires fulfilled in this bee from Spain and Portugal.

The French bee

It is curious that with one exception the French bee has been accorded little notice outside its native land. Between the two world wars hundreds and thousands of artificial swarms found their way to England, and the importation of package bees goes on even now though to a less extent than previously. The outstanding ability of the French bee as a honey-gatherer is a recognised fact in England.

Some 30 years ago I imported a substantial number of artificial swarms from Southern France and I have an extensive experience of the local strains which I came across in the different parts of the country. Although all these local strains have basic characteristics in common, they show essential differences in the way they emphasise one or other of these characteristics.

There can hardly be any doubt that it is a race in its own right, developed since the Ice Age from the Iberian bee as another variety of the Tellian group of races. All the characteristics of the Tellian are seen here as in a mirror. The rough edges of the character of the North African bee are more polished than in the Iberian, which is the intermediate race, with the one exception of the tendency to sting. On the other hand, a number of traits appear, developments of course of potentialities that were already present in the primary race, of real economic importance.

The inordinate aggressiveness and fierceness of temper is the sole cause of her bad reputation in England and it is without doubt also her most serious drawback. It is not just a question of a strongly marked tendency to sting, but an urge to attack without reason and provocation anything that is in neighbourhood of the hive. Although this urge is a marked characteristic of the whole Tellian group it reaches its maximum, as far as I know, in the bees of Southern France. On occasion this extreme aggressiveness can constitute a real menace.

On the other hand, the French bee is endowed with an amazing energy and capacity for work. Indeed, as my evaluations have again and again shown, she clearly embodies the maximum ability for production of honey of the whole Tellian group. Her performance on the heather, especially in a second-cross, is unsurpassed by any other race.

The Tellian and its Iberian sub-variety makes very watery cappings without any space between the honey and cappings. This kind of capping is of no consequence where extracted honey is produced, but spoils the appearance of sections. Among the French bees we came across strains with white cappings, however, not of the perfection of finish as made by the old English native bee; that is, pearl-white, raised and dome-shaped, and the outline of each cell clearly visible.

The strains in Southern France are generally very prolific with a loosely arranged brood pattern. In the northern half of the country, there is a progressive falling off in fecundity and a tendency to a more compact brood pattern. The swarming tendency is very much more pronounced in the southern strains than those of the north. As the honeybee does not recognise man-made demarcations and national frontiers, we can only consider strains of a given locality or region.

As with all varieties of the Tellian, the French bees have a very marked hereditary susceptibility to brood diseases. All the West European varieties are afflicted with this hereditary defect and in my view is a racial characteristic of the Tellian group. The old English bee was no exception, and as long as she was about our apiaries were never free from both kinds of foul brood, sac brood, chalk brood and a number of genetic abnormalities. Queens laying infertile eggs were common – a defect which to my knowledge does not occur in any other race or groups of races. It was curious that when queens and bees of this group of races were again introduced, sac brood and the other abnormalities quickly reappeared. A seeming disregard for the infestations of wax moth is another serious weakness of this group of races.

Although at first sight the French bee would seem to be quite unsuited for cross-breeding purposes she is, in fact, perhaps more eminently suited for it than any other race. She is endowed by nature with the whole gambit of all the good and bad traits of the Tellian, but no longer in their original intractable form. My experiments have shown that it is easy to breed out the worst defects, such as bad temper and tendency to swarm. Although there are other races which have greater potentialities for higher production than the French bee, yet this variety possesses a unique linking of the most economic factors in an intensity not found in any other race or group of races. Moreover, this bee also offers us in the most favourable form for breeding purposes all the latent possibilities incorporated in the most important group of races. We cannot, of course, ignore the many very undesirable characteristics, but these offer no insuperable obstacles. I will give an example from my experience to show the possibilities this race can offer and the surprising results that await us in cross-breeding and the synthesisation of new combinations.

As we know, the French bee is black or dark brown, very aggressive, and given to stinging, extremely nervous and prone to swarm, propolises badly, and is highly susceptible to every known disease and abnormality of the brood as well as acarine. Yet in spite of this series of bad characteristics, we were able within a period of seven years, to develop from a cross with our own strain, a new type which in colour was a deep golden, a golden tint which was far more attractive and striking than any other golden bee which has so far come to my notice. What was of greater importance, this new bee virtually could not be provoked to sting, and showed itself more gentle than the gentlest of Caucasians. It was, moreover, very quiet in behaviour, betraying not the least trace of nervousness when manipulated, did not swarm or propolise, was very prolific, excellent in performance, and showed no sign of any brood abnormality. In place therefore of a series of highly undesirable characteristics we had evolved a bee with an exactly opposite set of the most desirable characteristics, and this to a degree never previously brought about – and this from stock which seemed in every way the most unpromising. Unfortunately, this new bee had one major drawback: it was extremely susceptible to acarine, a defect apparent all along in the French stock and which again came to light in a highly accentuated form in the new type.

For commercial purposes the French bee responds best when crossed with Italian or possibly Greek drones. However, we have here a classic example of an instance where the best results for honey production are achieved in a second-cross, especially with a back-cross to drones of the above-mentioned races. On the other hand, Italian queens crossed with French drones give best results in a first-cross.

The Nigra

As I have already pointed out, the further north one goes in France the more noticeable a progressive gradation in the harsher racial traits of the typical French bee. In place of the exceptional fecundity and loose brood pattern, we find a more limited laying power and a compact brood-nest. In the Swiss 'Nigra' we have these developments at their fullest extent, with the exception of colour. In the Nigra the pitch black of the Tellian reappears in a completely undiluted form.

The Nigra represents the classic form of the mid-European bee as regards the qualities which bear directly on economic usefulness – fecundity, industry, swarming propensity, temper and aggressiveness, wintering ability and hardiness, comb-building power, provision of reserves of pollen and honey and, above all, in an orderly arrangement of the brood-nest. Furthermore, in this race certain peculiarities appear which I have never seen anywhere so emphasised as here, two opposite characterstics for which the Swiss have coined two special words, 'Hüngler' and 'Brüter'. The first type will concentrate its entire energy to the gathering of honey when there is a flow to the neglect of brood-rearing; the other kind tends to turn all the newly-gathered honey into brood – a characteristic for which the Italian bee is noted.

We have given this bee extensive trials and are fully aware of its economic value and breeding potential. Unfortunately, in our honey-flow conditions, the Nigra displays an unmanageable swarming disposition, particularly in a first-cross, with the result that this bee has proved economically useless. In order to ascertain whether this extreme proclivity to swarm was due exclusively to environment, further experiments were made in a commercial apiary some 125 miles from Buckfast, but the results were exactly the same. However, first-generation hybrid queens crossed back to Buckfast drones gave us highly productive colonies with little swarming. This is another example of the fact that one cannot rely on a first-cross for the most favourable effects of heterosis and the best economic performance.

The old English bee

The old English brown bee, another branch of the Tellian race group, has often been referred to in these reports. She lives today only in the memory. Some 50 years ago she fell a victim to the acarine epidemic and was completely wiped out. However, I consider it only right to record here some of her outstanding characteristics, for although it has now no direct value yet it can help us to build up a picture of the ideal we should be aiming at in our breeding.

This dark brown bee was the possessor of quite an extraordinary assembly of most valuable economic qualities, but in a much more disciplined form than found in her two nearest relatives, the Nigra and French bee. The main difference between the English and the other two was that the former had a very restricted fecundity. The maximum brood area of a colony of this race scarcely ever exceeded eight combs of British Standard size, i.e. $14 \times 8\frac{1}{2}$ in. The disadvantages of such a low fecundity were, however, in great part counter-balanced by a very unusual longevity, wing power and industry. She was thus well adapted to the prevalent climatic conditions of the British Isles and so guaranteed survival with a minimum of assistance. Although she did not give us the crops we expect as an average today, she was a bee that for practical purposes did not require feeding and proved self-supporting. These traits, extreme thrift, ability to fend for herself, longevity, hardiness, wing power and industry, which were such a marked feature of the English bee, can hardly be found together in the same concentration in any other race.

I have mentioned the abnormal wing power of this bee. This is not a mere hearsay report, but one I can vouch for from my own experience. Before 1916 we regularly had a crop of heather honey without moving our bees to the moor, but since the extinction of the English bee that has not happened. In 1915, to give an example, the bees at our home apiary averaged close to 1 cwt. per colony, including winter stores, from the heather. The nearest heather was some 2¾ miles away and at a height of about 1,200 ft. it must be assumed the bees had to fly a further mile or two into the moor, or a total distance of no less than 3½ miles. Experience has shown that this exceptional wing power is a mark of the Tellian race group.

The English bee had two other qualities which make me regard her as an ideal: her incomparable cappings and her capacity for building comb. I have already given details of her cappings, but her ability to build comb was no less remarkable. It was quite extraordinary to see the zest and speed with which this bee, even in the poorest of flows, built comb and drew out foundation to a pitch of perfection as otherwise seldom seen. Indeed, in regard to these two traits there is no other strain or race that can match her. We had the good fortune to be able to retain the latter trait in our strains, although not to the same degree of perfection.

As is well known the old English bee suffered from one serious defect which eventually proved fatal to her, her extreme susceptibility to acarine. The result was that in the space of a mere 12 years the whole race was exterminated by this disease. For many years a hope was entertained that perhaps a few colones might have survived in such remote places as the Outer Hebrides, but all efforts to find any have proved in vain.

From the immediate economic point of view the loss of this race was not irretrievable, although about this there are great differences of opinion. But one thing is certain, the loss was irretrievable from the point of view of the latent possibilities in this race for cross-breeding and the development of new combinations. Our own strain is a good example of what could have been done here, since it is derived from a cross between the Italian, as imported 50 years ago, and the old English drones.

The Heath bee

This bee, known in England as the 'Dutch bee', is a further development, or perhaps it would be truer to say, a reversion to the parent stock of this group of races. The Heath bee is commonly regarded as a special race mainly because of her extreme proclivity to swarming. This proclivity has been encouraged over a long period of years by the professional beekeepers of the Lüneburger Heide who, with the help of a highly specialised management, have fostered and turned this swarming propensity to the best possible advantage for a flow in August from the heather which forms their main crop. The extraordinary vitality and swarming proclivity and all the other characteristics of the Heath bee spring from her Tellian inheritance. The Heath bee is not found much outside the Lüneburger Heide and the adjoining Holland, and she can have little claim to real value where there is no late flow and a corresponding method of beekeeping.

This bee, however, played a very important role in the re-populating of the country with bees after the acarine epidemic. A great number of colonies in skeps were imported for this purpose from Holland by the Ministry of Agriculture in the years following the First World War. The extreme swarming tendency was of great service also in this instance. But apart from its advantages in the circumstances mentioned, this race has not found favour.

North-East European and North Asian sub-varieties

The Tellian race group, as I have already said, extends right across North-East Europe and the Northern half of Asia. Within the confines of this immense land mass with all its variations of rolling steppe and dense forest and its extremes of heat and cold, there must be an untold number of sub-varieties of the Tellian bee, endowed above all with an ability to withstand exceptional cold and very long periods of confinement in winter. Unfortunately, on this point we have no first-hand information, but doubtless many surprises lie in store for those whose search leads them into this vast expanse.

The finnish bee

It stands to reason that in the far north, south of the Arctic Circle, between the Atlantic and Pacific oceans, varieties of the honeybee will be found that are able to resist extreme wintery conditions and confinement many months on end without the possibility of a flight. A tolerance of extremely low temperatures is probably less vital than an ability to endure long periods of confinement without a cleansing flight. This ability in turn presupposes a host of other traits. A high degree of longevity, an intense quiescent disposition, an ability to subsist on a minimum of food, including a capacity to survive on stores of doubtful quality, are clearly essential contributory factors. At the same time the short summers, limited to a brief few months, presupposes likewise the ability of a rapid build-up of colonies and a corresponding tendency to swarm, to make good the inevitable winter losses. In any case, a colony not in the best possible condition by the autumn would have no chance of survival in the arctic conditions. Natural selection and its elimination of the unfit insures in conditions of this kind a progressive intensification and preservation of the qualities survival calls for in a relentless and brutal way.

At the outset of our cross-breeding experiments we were fully aware of the difficulties we would meet in isolating the qualities enumerated in our cross-breeding experiments. The

Finnish bee is a distant branch of the Intermissa. So also is the Swedish variety. As we feared, all the undesirable traits of the Intermissa would inevitably come to the surface in a more intensive form than in the prototype, due to the relentless struggle for survival in the fiercest of climatic conditions. We made the initial crosses in 1968, but full success in synthetising a new combination, incorporating the particular desirable qualities of the Finnish bee, has up to now eluded us.

APIS MELLIFERA SAHARIENSIS

We come, finally, to a very interesting race indeed, not only from the point of view of its origin, but also from that of its economic and breeding possibilities. Only a few years ago the very existence of the race was held in doubt, and the origins and descent of the Saharan bee will probably for ever remain a mystery. She differs in so many respects from every other race of the honeybee, while in her external markings and general behaviour she comes closest to the Indian bee (*Apis indica*).

Our evaluations have shown that the pure Saharan is not particularly prolific. In her behaviour she is quick and nervous, but in spite of this nervous temperament I would not describe her as bad tempered. If unsuitably crossed, however, they can be very ill-tempered. Her honey-gathering ability has led some to suppose that she has an abnormal tongue-reach, but this has been shown to be incorrect. She has great wing power and unusual industry in foraging together with a measure of hardiness and ability to winter in our northern latitude. But she does tend to fly out during the winter when weather conditions are unfavourable with the consequence that the loss of bees is high. The cappings of this bee are grey to very grey, and she constructs brace-comb and uses propolis, but in moderation.

As far as I have been able to ascertain the pure Saharan bee has no real economic value, at least in temperate climates. When suitably crossed she shows great promise. Mere chance crosses are, however, completely ruled out here. A first-cross Saharan queen with Buckfast drones gave truly amazing results in performance. The seasons of 1963 and 1965 were total failures, and the weather was indeed so bad that in both years no attempt was made to transport the hives to the heather, the only time this has happened since 1924. The year 1964 was an average good year with a production of 81½ lb. per colony.

The Saharan first-cross, however, averaged 231 lb. The winter stores amounted to a further 23¾ lb. per colony, which was 3¾ lb. less than the general average. This quite outstanding performance was largely due to the phenomenal strength of the colonies, and partly to the vitality, longevity, wing power and industry of this first-cross. The strength of the colonies reached such a pitch that many of the types of hive in common use would have proved completely inadequate.

The details about fecundity, the compactness of the broodnest, the rearing of drones, and the absence of any tendency to swarm, which were mentioned in the report of my journey to the Sahara, have all been confirmed by further experience. In the meantime, I have come across an additional noteworthy charactersitic, an extremely active capacity for building comb which, of course, is an essential concomitant of an outstanding honey-gathering ability and an absence of swarming. Moreover, it is not merely that they draw out foundation with amazing rapidity, but the combs are completely faultless with a minimum of drone cells.

Unfortunately, the Saharan suffers from two rather serious defects: she is very susceptible to paralysis, and appears unable to ripen heather honey properly. This latter defect is not one which appears in every colony, so that there is ground for supposing that with careful selection those negative traits can be bred out. Indeed, the further breeding results have fully confirmed this supposition.

Supplementary findings

The additional tests, made between the strains of the various oases, have shown that there is no notable difference in the characteristics in the bees of one oasis and that of another, apart from the slight nuances that make their appearance in every race.

Indeed, the Sahariensis is marked by a great uniformity, similar to the Egyptian and Cyprian races. The particulars indicated in the foregoing section still hold good, apart from a few exceptions. No special susceptibility to paralysis manifested itself in the newly-acquired strains. However, some showed an increased aggression – again a variation every race will exhibit, even in the case of those deemed unusually gentle. As already pointed out, in their native habitat we were able to pull the colonies apart without any protection. As to their fecundity, the further results obtained, if anything, surpassed those secured in our original comparative tests. There is no doubt a Sahariensis cross, suitably matched, will achieve a colony strength surpassing that of any other race or cross. But it is imperative that the beekeeper who strives to obtain the best possible results from a cross of this kind uses a brood chamber of adequate capacity and the needed extra super space. Suitably paired crosses are, in addition, of paramount importance. Where no control of the drones is possible, to ensure select matings, no attempt should made with this bee. According to our experience, the real worth of the Sahariensis resides in its breeding potentialities.

The origin of this race presented at one time a seemingly insoluble conundrum. However, the biometric findings of Prof. Ruttner indicated that this race is a far off sub-variety of *Apis mellifera adansonii*. The fact that a large part of the Sahara was at one time, during the Ice Age, covered in vegetation supports these findings. Our own observations, regarding the physiological characteristics and the general behaviour of the Sahariensis, substantiate the conclusions drawn from the biometric findings. The mysterious origin of *Apis mellifera sahariensis* may therefore now be regarded as solved.

CONCLUSION

With these research journeys, the evaluation of the different races and testing of the experimental crosses, we have taken the first steps in a project whose aim is to utilise the possibilities which Nature offers us in the various races of bees. The realisation of these possibilities, with the help of the latest breeding methods, will enable us to open up to beekeepers the kind of advantages up-to-date beekeeping imperatively demands.

As we can see, Nature with the means at her disposal has in no way produced a 'best bee' or an 'ideal bee', still less a race of bees which answers all the desire and needs of the modern beekeeper. The results of evaluating the different races makes one thing clear: every race has its advantages and its drawbacks, its good and its bad characteristics linked together and emphasised in a host of different ways, which have been determined arbitrarily by environment and chance. Likewise, each race comprises a number of good and indifferent strains, with by far the majority in the latter category.

Breeding experiments up to the present have been confined to the improvement and intensification of uniformity of particular races, but these will never be adequate to meet the demands of the future. True the efforts made are of undeniable economic value, but at the same time the possibilities are clearly limited. Inbreeding brings about in the honeybee a serious deterioration in vitality which raises unsurmountable problems in many directions.

The synthesisation of new combinations by way of crossbreeding is, indeed, the only breeding worthy of the name. It alone enables us to actualise all the different potentialities involved. For it alone has the power of bringing together all the various races and strains with their desirable economic qualities, of combining these into new types of bee, while at the same time eliminating the deleterious traits, and thus producing a bee which will completely answer all the needs of modern beekeeping.

19.50